AMERICAN GARDENS OF THE
NINETEENTH CENTURY

American Gardens of the Nineteenth Century

"For Comfort and Affluence"

ANN   LEIGHTON

The University of Massachusetts Press

Amherst

Library of Congress Cataloging-in-Publication Data

Leighton, Ann, 1902?–1985.
American gardens of the nineteenth century.
Bibliography: p.
Includes index.
1. Gardening—United States—History—19th century.
2. Gardens, American—History—19th century.   3. Plants,
Cultivated—United States—History—19th century.
I. Title.
SB451.3.L46   1987   635'.0973   86-11330
ISBN 0-87023-532-X (alk. paper)
ISBN 0-87023-533-8 (pbk. : alk. paper)

To my sister, Emily,
and my daughter, Emily,
and to my daughters-in-law, Jane and Ellen.

Contents

Laying on of Hands

Domestic Pleasures

Contents | ix

List of Illustrations

Foreword

With every succeeding book on the same subject, the list of those to whom each book owes much of its existence grows. The same libraries and societies again pour out their treasures; the same faithful relatives and friends endure all; the same doctors, augmented by ever younger successors, perform similar miracles of restoration and prolongation, aided by ever younger and more helpful nurses; enduring typists survive to type again from unimproved handwriting. So an author of a third book must simply bow again gratefully to the entire company of those who helped through explorations of the gardens of the first two American centuries, and then must add the new guiding lanterns from her voyage through the most difficult and confused of all centuries— the nineteenth.

One of the joys of research is finding unexpected treasures on the way to determined goals. As I sought for books to illuminate the seventeenth and eighteenth centuries (leaving no secondhand bookshop unentered), one solace for sacrifices made to obtain the earliest genuine sources was a harvest of later garden books, from the prolific nineteenth century. Although I had not considered going beyond the first two centuries, books on the nineteenth century were plentiful and cheap, and they held incredible adventure tales, charming absurdities, and an account of the burgeoning social awareness. These books caught my eye and filled my shelves.

This festive plenty became my workaday mainstay when the world's foremost horticultural library was suddenly decimated. The diminution of the superb Massachusetts Horticultural Society Library, with the sale of its treasures and the boxing up and removal of its working historical references, was a great loss. Thankful for my past blessings (like being able to check on Washington's library in the Boston Athenaeum, and then to work with the same editions ready to hand in the Massachusetts Horticultural Society Library), I can only marvel at my former good fortune, and congratulate myself on every spendthrift gesture. A saving grace appeared in the generosity of the Pennsylvania Horticultural Society, which cooperated sympathetically by mail. I bow in gratitude that there is still a library of its caliber.

To the Hunt Botanical Library belong particularly happy memories of books stacked high for a day's visit, with a rich assortment of references for Nuttall's sketches and wanderings.

The Essex Institute and the indefatigably resourceful Irene Norton were, as always, most amusingly rewarding, with elaborate examples of uplifting floral sentiments in lofty poetry and moral tales.

Among the individuals who may not have been sufficiently recognized previously—besides all my staunchest aides in my children and their families—I especially thank my son, Jim, and his wife, Ellen, for totally efficient advice and assistance in finally wrapping up this book and bearing with me throughout its long delays. To my sister, Emily Luce Pierce Niles, I owe a new debt, for late, generous companionship equal to our earliest childhood's shared adventures—still one of the most rewarding and entertaining of relationships.

To my son-in-law, Tom Cain, and my daughter, Emily Cain, I owe all my Canadian contacts and pleasures, including help from yet another superb horticultural library and librarian, Ina Vrugtman at the Royal Botanical Gardens in Hamilton. My grandson, Patrick, contributed his liking for historical research combined with cheerful companionship. And Nina Chapple, architectural historian with a special flair for Victorian charm, is to be thanked for the vigorous account of the garden practices of the Howland family in 1867–68.

A resourceful unit of helpers, the Tom Smith family, has continued its support, through the paper centuries, ever charming and willing, and still as addicted to garden touring and secondhand bookstores as at the very beginning.

To Pattie Hall, with whom I have shared a variety of tasks, I owe thanks for inestimable help both inside and out. As our instincts for where to place plants in a garden seem nearly identical, I dared to trust her with sorting out the comments of nineteenth-century experts, especially of authors Breck and Scott, to be of use today. She even came to share my fondness for them both.

Dorothy Monnelly, my nearest neighbor and a dedicated gardener, has been a cherished checker-in and an instant resource in emergencies for copying pictures from the shabby but heartwarming books.

To the infinitely professional, inspired photography of William Owens I owe the bulk of my illustrations, made by him on a very hot day in a stuffy dining room, surrounded by book treasures I would not let out of my sight.

My tried pen pal, Frank Coleman of Mountain Shoals Plantation at Enoree, South Carolina, has been unfailing in interest, information, and encouragement.

Heritage Plantation, in Sandwich, Massachusetts, contributed from its great collection of Currier and Ives.

My friend Charles Eliot helped me with details of the career of his uncle, Charles Eliot, as described in his grandfather Charles W. Eliot's biography of his son. The Portsmouth Athenaeum furnished me with a copy of this masterful book.

Rebecca Rogers, who lives in a house once occupied by Dr. Kirtland, has furnished a wealth of material about that remarkable individual.

Melanie Simo, our emerging American authority on Loudon, has been ever present with photocopied references, book loans and purchases, and an infinite sympathy for my enthusiasms.

And, as mercies never cease, at the last moment, when I discovered a book I had not known existed and instantly felt I needed, I called the bookseller and learned that I had missed it by minutes. Kindly Jim Hinck, of Anchor and Dolphin Books, arranged a loan with the lucky purchaser, June Hutchinson of the firm Private Gardens, so I could have a look at it. This book became the very welcome last-comer to my bibliography and proved that gardeners and booksellers share very special qualities.

Deborah Robson, whom I have not had the good fortune to meet in person, has ridden stylishly to the rescue and become the final arranger of all the treasures and trifles I have gathered with such pleasure.

To any who find my compliments too flowery, I can only reply that this sort of help makes everything bloom. Paramount in providing an essential service has been Jane Langton, the only expert who has never complained of my penmanship and to whom I owe gratitude for all the drudgery and kindly enthusiasm of getting into fair copy one more garden book.

And as a last gesture of gratitude in a race into print, I thank Theora Frisbee for helping to pick up, in Florida, all the final bits and pieces from their New England origins, and my sister Emily for managing all the complex logistics of manuscript and typescript.

AMERICAN GARDENS OF THE
NINETEENTH CENTURY

Preview from the

Summerhouse

S UMMERHOUSES were American garden fixtures from the time householders ceased to worry about being surprised by Indians and began to fancy a small sitting area near the house and in the garden, from which they could see all around. Originally made of lattices and often covered with vines, by the eighteenth century the summerhouse had become a grace note in American living. In the early eighteenth century, Michel Guillaume Jean de Crèvecoeur wrote enthusiastically of resting in one in William Byrd's garden.

Within the vagaries of the American climate, where even the coldest areas have periods of intense heat and sunbrightness, the summerhouse—not to be confused with the arbor, another fixture—afforded an outdoor sitting place, near the house and garden and orchard (all equally inevitable), from which one could see all around, and be seen by none if the vines flourished. On Maine farms, the summerhouse was resorted to on hot afternoons as a pleasant place to sit and shell peas. Nesting places for birds inside the eaves assured human sitters some protection from insects. Here visitors could be entertained, the men of the household could meet to talk and smoke, and the ladies— in their own time—could gather to sew and chat or read.

Vines from remote territories (like Ohio) were favored for the cover-

Summerhouse. One of A. J. Downing's suggested "embellishments"
from his *Landscape Gardening*. A "covered seat or rustic arbor with
a thatched roof of straw. Twelve posts are set securely in the
ground, which make the frame of this structure, the opening between
being filled in with branches (about 3 in. in diameter) of different trees,
the more irregular the better, so that the perpendicular surface of
the exterior and interior is kept nearly equal. . . . The figure
represents the structure as being formed around a tree. . . . The
seats are in the interior."

ing of summerhouses, to make denser shade. Especially spectacular
blooms were fancied to tempt hummingbirds, which—according to
one visitor—"fanned one's face with their tiny wings." The style of
the summerhouse echoed prevailing tastes. In this unusually volatile
century, a small classical temple or a Moorish mosque could give way

to a rustic shelter with a thatched roof and plaster ceiling, into which the ladies of the household were encouraged to stick shells on rainy afternoons.

Gardens, from their colonial beginnings, had been blessings as well as necessities, as inevitable as roofs, fences, and wells. Expected and accepted by early travelers, they bloom in old letters and accounts. Books on how to plant, use, and maintain them stood beside family Bibles. Apothecaries could assume their existence and depend upon them. Recipes for cooking meat, fish, and pies show what had to be available in them. Linens were scented and blankets kept from moths by pungent leaves and flowers harvested from the garden. People walked in their gardens, and sat in them, and built outhouses and toolhouses and summerhouses and wellhouses and birdhouses in them.

The early nineteenth century was not a time for Americans to pause and ponder trends in garden design. Conflicts abounded; the success of a wholly new sort of government swung in the balance. Disaster loomed over the growing problem of who and how many were actually free, truly to be counted as citizens. Nothing was sure but the will and integrity of the now disappearing founders. Discussions of the relative merits of use, beauty, and truth in horticulture and of the purposes of Abigail Adams's "nature's handmaiden, art," were for older countries across the Atlantic. Edmund Burke, who had done his eloquent best for the colonies, would be remembered in excerpts from his political speeches but not by his written dissertations on beauty. Pronouncements on taste by Sir Uvedale Price, widely noted by secure English landholders, would reach American gardeners belatedly and chiefly at third hand.

Washington had left his legacy to gardening and agriculture in his careful planning of the gardens and grounds of Mount Vernon and in records of his experimental farm. His pleas for diversified crops on improved land, as an alternative to the ruinous one-crop tobacco syndrome, would echo through the century. Franklin had left the seeds of special societies and public libraries to further the knowledge and practice of botany and agriculture. Remembered today in Paris for his foresight, he advocated planting cities with street trees for the benefit of public health.

To Jefferson, gardeners across the world owe a special and lasting debt. With the Louisiana Purchase in 1803, he had peacefully doubled the size of the United States. Explorers he sent out to find a route to the Pacific brought back plants to enrich American gardens and become collectors' joys in the old countries. At the same time, the little gardens on the American seaboard would be packed up, root, seed, and scion, and slowly transported westward to sustain and cheer, comfort and cure their owners, as they had done their owners' ancestors in both the New World and the Old.

Knowledge of the rich soil of the Ohio country had long intrigued early settlers of the coastal regions, who were irked by British reluctance to protect venturers beyond the line of mountains that formed a natural bulwark between the developing colonies and those unexplored lands. Jefferson's Louisiana Purchase lifted the barriers and spread out a wide, if wild, carpet for exploration and development.

As if this were not enough to occupy a young government, there was also an urge to press on, to cross the entire continent and establish a route to the shores of the Pacific Ocean. It seemed obvious to those at home that the feat could be accomplished by linking up great waterways running in opposite directions with short portages over passes of what was assumed to be a single, though very high, western range.

Planned and instructed by Jefferson, the Lewis and Clark expedition was the first of the opening wedges. Early fur traders, pressing in from both sides of the continent, had established rudimentary lines of communication, frail meshes of meetings and markings, upon which later eager botanists would depend. On this divergent group of individuals, some sprung from our own eastern shores and some arrived from a curious and acquisitive older world, fell the unasked-for responsibility for furnishing information far beyond their plant specimens. Useful they could be; used they often were.

When an era of unleashed territorial expansion followed the first tentative explorations, with the pushing back and putting away of the Indians, the constant controversy over states' rights to declare themselves slave or free, and the wheeling and dealing that kept the whole country in turmoil, it is no wonder that the word *contrast* commanded first place with gardeners in their floral and arboreal layouts.

Beyond views from any New World summerhouse, the nineteenth century around the world was awash with revolutions—social, philosophical, scientific, and industrial. News of turmoils abroad arrived

Illustration from seed catalogue ("Gardening," frontispiece from Washburn & Co., *Amateur Cultivator's Guide,* 1869).

slowly in the United States. Americans were preoccupied with the development of steam, gas, and iron, and with their own social upheavals, from the blood-soaked abolishment of slave labor to the importation of hordes of untrained labor from overcrowded lands abroad.

One of the gentler modifications of society in the American nineteenth-century world would be the special attention paid to women. Women had not before been so openly deferred to as presences in their own right. In the early nineteenth-century family, the mother reigned as guide, tutor, and moving spirit. The father was the provider and the final authority, but the mother made all the more complicated judgments. Not surprisingly, women would become the teachers in the new free schools across the country. The seventeenth century (when Indians accused English settlers of spoiling good labor in their wives) had shown women respect as property owners. The eighteenth century had allowed women to become ornamental. The nineteenth-century role of social arbiter was confining for these women, although it may have seemed glamorous to their sisters abroad, where a few individual English women were beginning to assert themselves in letters and government.

One of the great contributions of a democratic America in the nineteenth century was the establishment of a system of universal education in publicly funded schools. The institution of compulsory education directly influenced gardening and farming in two ways. First, botany was introduced as a required study for schoolchildren. Second, a market was created for how-to-do-it books on floriculture, agriculture, landscaping, and village improvement.

Another movement, more or less original, was for the establishment of public parks with a concomitant attention (here almost totally American) to suitable systems for the burial of the dead and for the maintenance of cemeteries, which should also be of advantage to the living. This revealed the most obvious impact of "natural," less geometric, landscape gardening, the so-called English School, which was beginning to blot out many of the historic formal gardens of the Old World.

In the midst of this flurry, a whole new concept arose, originating with the early social economists in England, of the greatest good for the greatest number. Associated with this was an idea, nurtured in American churches since Puritan times, of a debt owed the less fortunate, including the agonizingly poor and overworked. Lady Bountiful would later become a figure of fun among those relying on public support, but she was a brave figure for her times.

And they were hard times, for the machine age coincided with the nineteenth century and quite literally twisted it in its own furnaces. The miracle is that gardening prospered, that in England, especially, the workers came home to their little plots and with the help of glass and frames began to raise hybrids of ordinary plants, which suddenly enlivened the horticultural shows and sent small plants with high-sounding familiar names to flourish in the gardens of dukes and in the public park parterres of all Europe.

All the while, the money from the furnaces and from the mines under great estates financed explorations in the most remote parts of the world to bring back horticultural and botanical wonders from high mountains, dank jungles, hot rocky islands, and steeply forested slopes. Gardens were truly involved with the developing history of the turbulent nineteenth century, with its deepest needs and its most exotic pleasures.

As travel became increasingly available to well-to-do seekers of cultural stimulation, palatial gardens could be copied exactly—though

"Budding's mowing machine is an admirable contrivance for cropping, or shearing lawns" (from J. C. Loudon, *Encyclopaedia of Gardening*, edited by Mrs. Loudon, 1871 edition): The two new moving spirits of nineteenth-century gardens—the lawnmower and the advancement of gardening ladies.

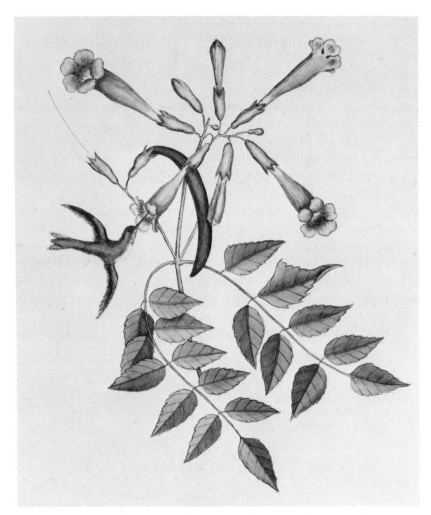

Trumpet vine (*Bignonia radicans*), a popular cover for summerhouses.

with local plant materials—and set down upon prairies or in Florida groves. As American architects began to feel free to collect motifs to suit their fancies and their clients' fortunes, those with the means could order up a small Alhambra, a modified Hampton Court, a Gothic cottage. The feeling of being so modern as to be heirs to all the ages was reflected also in gardens.

As the century wore on, an odd new manifestation of the growing

American character went beyond frontier trading instincts, and beyond the eighteenth-century gardeners' polite envy of their neighbors' unusual plants. It became important to impress others—even the total stranger. For this ephemeral and indirect pleasure, householders in the new suburbs were ready to sacrifice their front lawns, to groom them and leave them unused, merely to solicit the admiring glance of a passerby. Whole streets would be laid open, with long, untroubled vistas of green grass on either side, the houses set back uniformly from the street or sidewalk, their owners never publicly visible, represented only by this gesture of perfect, joining lawns for public view.

Another new idea for a young country was the concept of conservation of natural resources. Starting as a concern of a few individuals in the early part of the century, it grew to a universally important force by the end of the century with, by then, demonstrable disasters and advantages.

In the midst of all this, the individual's right to a place—a house and garden designed and maintained and enjoyed in the style of the times—would be central to the country's development.

Appropriately, the summerhouse we are taking for our model, from which to try to see the century whole, is our American summerhouse, not the later Victorian gazebo from which one was supposed to see all around. Ours is Downing's ideal summerhouse, seen by him in the Arnolds' gardens and much admired. Observing the outside world from these shelters, constructed primarily for seclusion and retirement, took serious effort. Distant views were minuscule. Nearby prospects were much broken up by the tangled timber of the structure, not to mention hanging vines.

Only this early American summerhouse, poised and protected, can furnish us the ideal vantage point from which to study the developing landscape of nineteenth-century America.

Unfamiliar Territory

ONE

American Plants and Early

American Botanists

ARDENS are living wit-
nesses of those who made them, tended them, discovered new plants
to go into them, and knew why each plant had to be there "for meate
and medicine, for use and for delight." Gardens cannot be separated
from their origins or their originators. To see a garden and not be able
to recognize its background or catch its figures of speech as it tells us
its history is like being at a party of strangers with no one introducing
guests to each other. If one is truly interested in gardens, a world is
open; but one cannot come in as a stranger and enjoy either oneself or
the gardens. Riots of bloom, intricate designs, plans so charming as to
be apparent under inches of snow or in driving rains—all these take
understanding. Copying without knowing why will miss the whole
point; polite pretense cannot avail.

One must seek to know the early owners and workers, the explorers
and the nurserymen, to listen to the visionaries, read their authorities,
learn the history behind each flowering plant. And knowing all the
plants' names is essential, besides being good garden manners—both
the familiar names and the scientific names with which scholars com-
municate around the world. William Coles, in his *Art of Simpling* of
1657, put it for us, describing the joys of gardening.

What a pleasant thing for a man whom the ignorant think to be
alone to have the plants speaking Greek and Latin to him, and

putting him in mind of stories which otherwise he would never think of. It will yield a man discourse wither soever he goes, travel he by sea or by land.

And in the end we will really know the gardens.

FIRST steps into American gardens from the beginning had to do with medicine. Food concerns could depend upon field crops, although seasonings soon followed medicinal plants into garden plots. Minor necessities like scents for close rooms, repellents for insects, tinctures for dyes, juices for scrubbing and shining metals—all these had to wait. The first, most important, plants were for reducing fevers, numbing pain, aiding in childbirth, soothing sore throats, expelling worms, making physics and tonics, staunching blood, and laying out the dead. Because syrups could cloak pure spirits and sweeten bitter doses, fragrant flowers like pinks and violets were included in the first raised beds in what were primarily gardens for health.

In the seventeenth century, precious medicinal remedies were dispensed by clergymen, government officials, and the leading philanthropic householders. By the eighteenth century, settlers had discovered more of their surroundings, were in better communication with each other and with the Indians, had interests and requests from abroad to fill, and were in a position to execute some authority in caring for others. The old standbys of the early housewife's gardens were then incorporated into the borders of handsome vegetable gardens, large enough to be walked in and shown off to interested guests.

By the nineteenth century, medicine, especially of the home variety, had become a modest but lasting amateur science. Books were written about newly discovered cures. Useful medicinal plants were listed in books about gardening and in seedsmen's catalogues. But they were planted to flourish with the vegetables, leaving "pleasure" gardens to follow the styles of the changing times.

Since the Middle Ages, the study of plants, which later became the science of botany, had been required of every man who wished to become an apothecary and perhaps later a doctor. Although the old herbals had endeavored to dispense wisdom with alarmingly look-alike woodblocks of plants found in medieval Germany (hoped to be the same as those advised by early Greek physicians), there was a need for books of plants illustrated well enough to be useful to everyone. By the seventeenth century, these volumes, often very handsomely produced, came with the first settlers to North America to be applied to the rich plant life of the American wilderness. Again, not surprisingly, it was discovered that fresh studies and new drawings were needed instead.

Old World countries were waiting to see what the New World might produce, in remedies especially. After the boon of "Jesuits' bark" against malarial fevers was discovered in South America by Spaniards in the sixteenth century, it had been hoped that a remedy for another universal scourge, syphilis, might also be found. In North America sassafras trees were promoted to this end until early seventeenth-century shiploads sent to Europe and England proved valueless. But there was enough evidence of the efficacy of indigenous plants used medicinally by the Indians to encourage plant collections and analyses. In the eighteenth century, fulsome publications on how to be one's own doctor had begun to appear. By the nineteenth century, when both people and gardens started to move westward, where less favorable climates were encountered, the number of those ailing and ignorant increased. As did openings for quackery. The need for scientific studies of American plants became more critical.

The requirement for botany as a premedical study by such established Old World universities as Edinburgh was insurance of a leaning toward botany in the few trained practitioners in the New World. There was a growing interest in which native American plants might be useful—and perhaps very valuable—to the rest of the world. The Indians consulted were, in many cases, full of information about plants and practices, though it was often doubtful if the remedies could be proven specifically beneficial. In such places as Yale College, it became customary for students to offer theses on local plant material and its possible usefulness. And now, as we stand poised to enter their world, upon which our own became so dependent, it is imperative that we meet some of these great men personally.

It is a vital and varied group and one where we are beginning to find—perhaps unexpectedly, certainly often overlooked—the founders of American botany. Among the first of these was Manasseh Cutler

(1742–1823), who illustrated Cotton Mather's "divine conjunction" in that he was both a clergyman and a doctor. For historians, he also represents a linkage of the best qualities of the eighteenth century moving into the nineteenth century and proceeding, with the new settlers, toward new western lands.

Cutler was born on the border between Rhode Island and Connecticut and graduated from Yale in 1765 with "high honors" and "proficiency" in astronomy, meteorology, and botany. In 1768, after he received his master's degree from Yale, he studied for the ministry and in 1771 was ordained pastor of the church in Ipswich, Massachusetts. Traveling and preaching throughout New England without access to books on botany, he appealed to the Harvard College Library to aid him in his investigations of "the botanical characters of such trees and plants as may fall under my observation which are indigenous to this part of America and have not been described by botanists." His project would include

> a catalogue of those [plants] which are found growing here but have been found in other parts of the world and therefore need no botanical description; and of such as have been propagated here but are not the spontaneous production of the country. An attempt of this kind, which I am not sensible has yet been undertaken will be necessary to furnish materials for a Natural History of the country in which we are at present very deficient.

Although Cutler's plea to the Harvard library had been particularly to borrow "Dr. Hill's *Natural History*," his manuscript is interwoven with references to Dr. William Withering's "late ingenious and useful" publication, *The Botanical Arrangement of British Plants*.[1]

Cutler found time—in addition to participating as a chaplain in the Revolutionary War—to study medicine, "driven to the practice of physic" to support his family. By 1785 the manuscript of his "Catalogue" was ready to be shown to a Boston society, the American Academy of Sciences, to which he had been elected a member in 1781. Cutler's work is entitled *An Account of Some of the Vegetable Productions Naturally Growing in this part of America botanically arranged by the Reverend Manasseh Cutler* and is recognized as the first treatise on New England botany. It sets the style and standard for later works that would occupy the attention of nineteenth-century gardeners.

In his preface, Cutler comments upon the deplorable situation in "an infant country" where nature has been liberal in its productions but the study of natural history has been neglected.

> The cultivation of this branch of science will open to our view the treasures we possess unenjoyed, and must eventually tend to the security and welfare of our citizens, the extension of their commerce and the improvement of those arts which adorn and embellish life.

Perhaps, he says, botany has been neglected because of "the mistaken opinion of its inutility in common life."

Here he makes an interesting observation on a coming conflict between two systems of plant identification. One system, devised by Linnaeus, depended upon counting the sexual parts to arrange plants into classes. The other, based on Jussieu, arranged plants by natural similarities. Cutler quotes Linnaeus as admitting that the "virtues of plants" may be most "safely" determined by their "natural characters," because plants of the same natural class are in some measure similar; those of the same natural order have a still nearer affinity; and those of the same genus have seldom been found to differ in their medical virtues. Linnaeus was a doctor himself, and this is a handsome concession.

In spite of the enthusiasm for scientific inquiry, many old superstitions and prejudices still held strong. There was still an ancient inclination toward using a little dose of poison to dull pain, which explains the presence of *Aconitum napellus,* or monkshood, in so many old gardens. A popular conviction also held that anything disagreeable must be very good for one. And much was left undetermined. Seventeenth-century herbalists had never been able to decide if white roses were binding and red roses loosening, or vice versa, although a conserve of the hips of red roses was recommended for seasickness on long voyages. Manasseh Cutler's entry on the wild rose or dog rose clarifies these mysteries.

> The blossoms gathered before they expand, and dried, are astringent; but when full blown are purgative. This species is generally preferred for conserves. . . . The pulp of the berries beat up with sugar, makes the conserve of hepps of the London dispensatory.

In this first documentation of useful American plants, advice and hearsay were included like helpful hints—a practice that held through many subsequent publications. There were carefully included warnings: that persons of irritable habits are most likely to "receive" the poison of swamp sumac; that the flowers and seeds of violets are said to be mild laxatives; that the yellow violet is used by the Indians for boils and painful swellings, to ease the pain and produce suppuration; and that silkweed makes excellent wick yarn, affording a clearer light than cotton wicks and a less offensive odor.

The local sassafras, whose aromatic bark is substituted by people in the country for spice, had many other uses. Bedsteads made of this wood are reputedly never infested with bedbugs. It is said to be an excellent diuretic and diaphoretic, efficacious in obstructions of the viscera, cachexias, scorbutic complaints, and venereal disease. An infusion of the roots makes a "grateful drink." A pungent hot oil possesses the virtues of the wood. Considerable quantities, according to Cutler, were exported to Europe.

Indians, Cutler says, considered the witch hazel (*Hamamelis* Linn.) a valuable article. They apply the bark to tumors. A cataplasm of the inner rinds of the bark removes inflammations of the eyes. The bark chewed is astringent and leaves a lingering sweetish taste. "The specific qualities of this tree seem by no means to be accurately ascertained. It is probably possessed of very valuable qualities."

Our black currants with yellow blossoms and black berries, like those recommended for sore throats in England, answer very well here in cases of infected tonsils. The bark sweetened with honey makes a good gargle.

The wild verbena, a simpler's-joy with quadrangular stems, was used by surgeons of the American army as an emetic and expectorant. They found its operation "kind and beneficial."

The elder (*Sambucus* Linn.) represented an ancient remedy in the Old World, which settlers were happy to find growing also in the New. Its uses, mentioned by Cutler, come under the cover of "Dr. Withering observes" that:

the inner green bark makes an acrid purgative useful in small doses as a diuretic and in dropsies; sheep with the rot will eat the bark and young shoots and cure themselves; the inner bark and

leaves make cooling ointments; a decoction of the flowers promotes expectoration in pleurisies; fresh gathered, the flowers
loosen the belly; externally in fomentations they ease pain and in-
flammation; they will give a flavor to vinegar; a rob [*sic*] of the
berries promotes perspiration; the flowers kill turkeys and the
berries poison poultry; the fresh leaves laid around young cucum-
bers and melons preserve them against insect. And the green
leaves drive away mice.

We have another achievement of benefit to his fellow Americans to
be credited to Manasseh Cutler—his move in 1787 to form the Ohio
Company for purchasing land for a "western colony." He helped
procure a grant from Congress and drove out in a sulky, taking twenty-
nine days, to visit the country. His next expedition was by ox cart, to
found the town of Marietta. By the time he died in 1823, back again in
Hamilton, Massachusetts, no longer considered a part of Ipswich, he
had—as Oliver Wendell Holmes was to say many years later of men
who might be considered truly to have lived—"shared the action and
passion of his time."

Early in the nineteenth century, Philadelphia was recognized as the
chief source of scientific enquiries. One of the guiding lights in Philadelphia, Dr. Benjamin Smith Barton, presided over a wide and active
circle, some of whose members (like Dr. Schoepf and Dr. Muhlenberg)
were busy writing their own *materia medica* of this country. Dr. Barton
always intended to write such a book of his own, a definitive work on
medicinal plants of the United States, but ill health and a constant interest in the projects of others prevented him from realizing this ambition. The nearest he got was his "Collections for an Essay towards a
Materia Medica of the United States." His preface, addressed to his
medical students, puts the case succinctly.

> I have thus, Gentlemen, endeavored to present you with a speci-
> men, or rather rude outline, of an Essay towards a Materia Medica
> of the United States. . . . In the pursuit of one of the most dig-
> nified and most useful of all the sciences, you are placed in an
> extensive country, the productions of which have never been in-
> vestigated with accuracy, or with zeal. . . . In whatever part of
> this vast continent you may be placed, you will find an abundant
> field of new and interesting objects to reap in. The volume of na-

ture lies before you: it has hardly yet been opened: it has never been perused. But by your assistance, the knowledge of the natural productions of our country may be greatly extended. . . . there are some of you for whom medical discoveries of importance are reserved—discoveries which would add a lustre to your names whilst they would ensure to you that which is more to be desired . . . an happiness that is imbosomed in the happiness of one's country, and the world.

Dr. Barton divided his "Collections" into nine headings: Astringents, Tonics, Stimulants, Errhines, Sialogogics, Emetics, Cathartics, Diuretics, Anthelmintics.

Of the first two categories, we grow several members for ornament today. Astringent plants found in contemporary gardens include *Geranium maculatum* (spotted cranesbill, whose root boiled in milk was used for cholera in children), *G. robertianum* (used in nephritis), *Heuchera americana* (American sanicle or alumroot, used by the Indians against cancer and as beneficial in ulcers), *Actaea racemosa* (black snakeroot or "squawroot," a cure for the itch and murrain in cattle; the roots make a gargle), *Uva ursi* (a valuable medicine in cases of "old" gonorrhea and nephritis), *Liquidambar asplenifolium* (sweet fern, used in treating diarrhea).

Barton offers a number of familiar plants as tonics. American native oaks are all useful substitutes for Peruvian bark, and *Quercus rubra montana* (Spanish oak) was used against gangrene. Plant materials used in fevers included *Prunus virginica* (wild plum), *Laurus sassafras*, *Diospyros virginiana* (the persimmon), and the barks of *Magnolia acuminata* and *M. glauca* and of *Liriodendron* (the tulip tree, also used as a stomachic for horses). Of the dogwoods, *Cornus florida* was long used in intermittent human fevers and for a fever called "yellow water" in horses. Barton used an infusion of the flower in a tea. *Aristolochia serpentaria* is offered as one of the more stimulating tonic bitters.

And that must take care of the hardiest among us, unless we may add that an errhine produces a nasal discharge, and the best is the well-known tobacco.

Another eminent physician of the early nineteenth century was Dr. Jacob Bigelow of Boston, whom we meet in another chapter. In spite

Swamp dogwood (*Cornus sericea*), drawn from nature by
W. P. C. Barton, nephew of Dr. B. S. Barton for *Barton's
Medical Botany,* 1817.

of several other calls upon his talents, he managed to write the definitive volumes on American *materia medica* of the Boston area, which he
later expanded to include New England. He studied in Philadelphia
with Dr. William P. C. Barton, a nephew of Dr. Benjamin Smith
Barton. Bigelow began his writing career with *Florula Bostoniensis,*
which he wrote while riding out to make calls. In 1817 he published an
American Medical Botany, being a Collection of the Native Medicinal Plants

of the United States. He used the Linnaean method of identification and illustrated his three volumes with "figures . . . engraved and coloured from original drawings made principally by myself."

Feeling that the "great body of medicinal agents now in the hands of physicians" needed no increase in number and might well benefit from reduction, Bigelow presented only twenty plants in each of the three volumes, "towards an investigation of the real properties of our most interesting plants." He avoided "exaggerated accounts of virtues," and if, "from a desire of avoiding error," he was not able to "establish fully the character of a native vegetable," he begged readers to recollect "that many foreign drugs which had been for centuries in use have still an unsettled reputation as to their powers and modes of operating."[2]

Last of the great early American doctors considered responsible for shaping nineteenth-century America and its gardens is Dr. Jared Potter Kirtland (1793–1877), born in Wallingford, Connecticut, the eldest son of Turhand and Polly Potter Kirtland. His maternal grandfather was the illustrious Dr. Jared Potter, surgeon in 1775 to the First Connecticut Regiment, physician, agriculturist, naturalist, and scholar. Jared's father was agent for, and one of forty-eight stockholders of, the Connecticut Land Company, which had purchased from the state of Connecticut in 1775 the tract of rich land known as the Western Reserve. In 1803 Turhand Kirtland took his family, including his brother and sister and their spouses, and all the family's children—except the ten-year-old Jared—to settle in Poland, Ohio.

Jared remained with his grandfather to be educated. No bright boy could have been left in more improving company. In addition to having a large library and a demanding practice, Dr. Potter maintained an experimental orchard, conducted a project for raising silkworms and keeping bees, and made a study of botany. The boy learned the Linnaean system from him, as well as the practical arts of budding, grafting, and recording.

At seventeen, Jared, having received a good classical education at the academies of Wallingford and Cheshire, went west to see his family, studying flora and fauna as he went. Sadly, at this time his grandfather died of a freak accident. Eating rye grains as he passed a field on his farm, he became infected by one that stuck in his throat. He died of gangrene. Young Jared Kirtland was left the library and funds to complete his education to become a doctor.

Kirtland continued his education in Wallingford and in Hartford. Then, as a member of Yale's first class to study medicine, beginning in 1813, he combined the study of botany with forays into geology and mineralogy and an independent exploration of zoology. The following year he moved to the Medical Department of the University of Pennsylvania to study, among other subjects, natural history and horticulture. He became a friend of Benjamin Smith Barton. In 1815 Kirtland returned to Yale to receive his medical degree, writing a thesis entitled "The Materia Medica Furnished by Our Indigenous Vegetation," unfortunately now lost.

He married and began to practice medicine in Wallingford. When his wife and child died of influenza in 1823, he resolved to give up medicine and to move to Ohio to help his father run the family interests. The need for doctors in Ohio was so great that he soon found himself with an extensive medical practice, stretching from Lake Erie to the Allegheny River. By 1825 he had built a house, married again, and lived on a farm of 244 acres with his wife and children, young nieces, several students, and the usual complement of hired men. He and his two brothers, who lived in equal prosperity nearby, took a particular interest in horticulture and the growing of fruit trees.

Evidence of Kirtland's accomplishments reflects a man of intense dedication to studies of natural phenomena. Even when serving three terms in the state legislature, Kirtland continued his research. He studied and wrote on mollusks and fish and assisted with the Geological Survey of Ohio, identifying and cataloguing multitudes of plants, animals, and birds. He formed the Kirtland Society of Natural History, which later became the Cleveland Museum of Natural History. Kirtland lectured at the Medical College in Cincinnati and helped organize the Medical Department of Western Reserve College in Cleveland (later the Case–Western Reserve Medical School). He was a founder of the Ohio Medical Convention. He helped found the American Association for the Advancement of Science and was appointed to the first Board of Managers of the Smithsonian Institution, where Andrew Jackson Downing, our first important landscape gardener, was later asked to lay out the grounds.

Kirtland wrote easily and generously on many subjects. He wrote regularly on scientific concerns and shared his views on home improvement and agricultural matters locally, through a magazine called

The Family Visitor. In an 1850 issue he expressed the country's indebtedness to Downing, whose influence, he said, could be seen in houses and gardens everywhere.

Dr. Kirtland represents the epitome of the sort of American who moved the country west.

IT would be remiss to end a consideration of American medicinal plants as the basis of American gardens solely with concerned doctors of the early part of the century. The devastating Civil War would seem to have had little impact on gardening literature, beyond effecting a pause in the promotion of new ideas and exchanges of new plants, but it created emergencies in medical supplies and agricultural produce. A nineteenth-century American plant book of historic importance emerged from the needs of the southern states. Dr. Francis Peyre Porcher was directed by the surgeon general of the Confederate States to investigate southern resources. He produced a small volume, which he later developed into a book (published in Charleston in 1869), *Resources of the Southern Fields and Forests; Being Also a Medical Botany of the Southern States.* In addition to recipes for "home made soda," from the ashes of saltwort (*Salsola kali* Linn.), and soap from soapwort (*Saponaria officinalis* Linn.), common early practices brought from the Old World, Porcher was careful to include Indian backing for such plants as Indian physic (*Gillenia trifoliata* Nutt.), which had been mentioned in the medical botanies of Bigelow and Barton. Porcher also quoted Rafinesque, an explorer whom we shall meet later, for instance in his recommendation of nine-bark (*Spiraea opulifolia* Linn.) in external applications for tumors, which Rafinesque must have learned from the Indians.

At the end of the century, in 1892, Dr. Charles F. Millspaugh, "physician, Botanist and artist," published two large volumes on *Medicinal Plants Indigenous to and Naturalised in the United States* and gave the "American aborigines" their due as early experimenters.[3] Arranged in the natural order are 1,000 plants, with their Indian references carefully quoted, as with goldenseal (*Hydrastis canadensis*), which the "American aborigines valued as a tonic stomachic, and application for sore eyes," and the Seneca snakeroot (*Polygala senega,* Dr. Tennent's much-publicized cure-all of the eighteenth century), and the wild cranesbill (*Geranium maculatum*), valued by the aborigines as an astringent in looseness of the bowels and "officinal," as noted by Dr. Millspaugh in the United States Pharmacopaeia of his time. "Officinal" originally meant "sold in shops" and is now the indication of a plant's listing in the current Pharmacopaeia. Many of the plants in this chapter are so listed today, though often with no mention of their original attributes and with little recognition of their history, which predated the European settlement.

That many of these plants appear in our gardens solely for our joy in them must justify our knowing their histories as cures for what ailed our ancestors.

NOTES TO CHAPTER I

1 An interesting aside here illustrates the circulation of plant materials and information among members of a young society of varied practitioners in the budding natural sciences and also demonstrates the void to be filled. Dr. Withering was the person who sent seeds of the foxglove, as a remedy in the treatment of dropsy, to Dr. Hall Jackson, of Portsmouth, New Hampshire. Jackson, in turn, sent the seeds to several physicians of note, including the Reverend Ezra Stiles, president of Yale College. In Philadelphia at the same time (1787), Caspar Wistar, anatomist, wrote to Humphry Marshall, botanist and author of the first book on American trees, to ask for leaves and seeds of the foxglove, although he mentioned that "a patient in the Edinburgh Infirmary who took the medicine as directed by Dr. Withering vomited to death." This nearly convinced Dr. Wistar to avoid using the medicine, although "dropsies are so often fatal that we must try everything." The foxglove's real value in treating that disease was to be discovered much later.

2 Because books of reference may be hard to find, some readers may be grateful anew to Dr. Bigelow for his list of the sixty plants he considered worthy. They were listed by him in a special index, by their familiar names. Following him, they are not alphabetically arranged here, and the spelling and spacing are as in the original.

Volume I. Thorn apple. Thoroughwort. Poke. Dragon root. Gold thread. Bearberry. Blood root. Cranesbill. Fever root. Poison sumach. Hemlock. American hemlock. Mountain laurel. Carolina pinkroot. Wild Ginger. Blue flag. Henbane. Bitter sweet. Indian tobacco. Sweet scented Golden rod.

Volume II. Winter green. Partridge berry. May apple. Skunk cabbage. Marsh rosemary. Butterfly weed. Small magnolia. Dogwood. Ginseng. Seneca snake root. Tulip tree. Butternut. American Hellebore. Blue gentian. Sassafras. Dogbane. Leather wood. Tall blackberry. American senna. Tobacco.

Volume III. Common Gillenia. Poison Ivy. Wax Myrtle. Common Juniper. Red Cedar. Black Bean. Bulbous Crowfoot. Starry Anise. Virginia Snakeroot. Star Grass. American Rosebay. Ipecacuanha Spurge. Large flowering Spurge. Bitter Polygala.

Sweet-scented Water Lily. Black Alder. American Centaury. Common Erythronium. Prickly Ash. Common Hop.

3   One of the special features advertised in the introduction of Millspaugh's book is the inclusion of "one hundred and eighty beautifully colored full-page plates: since the study of botany for medical remedies . . . without colored plates would be like the study of osteology without bones or the study of geography without maps." Dr. Millspaugh made these illustrations himself from "fresh living individual plants . . . aided by experienced botanists," so someone must have had an extensive garden and intimate experience in growing plants—resources and skills now sadly lacking in modern instruction on landscape gardening arts and their history.

Star of Empire: Botanist-Explorers
in the Western Territories

A word can change its meaning almost before our eyes, like a "florist's flower," blooming and then fading, to reappear less bright and be no longer sought after. *Empire* is a case in point. Once glamorous, it has now faded and is treated as a weed, scorned even in memory. In the early nineteenth century, empire builders were great men, empire a splendid dream.

The underlying force was the desire for land, of which there seemed to be an infinite supply. In such a rush for possession of riches, there are always seventh sons of seventh sons, who go out with their fates over their shoulders and discover wonderful things for themselves and by themselves. Botanists and other natural scientists in the nineteenth century belonged to this innocent class of adventurers, as they scavenged a countryside for its plants, shells, birds, rocks, and bones—overlooked as trifles by those set only upon enlarging their landholdings. Botanists in early nineteenth-century America, though remaining largely solitary in their ways, often had to rely upon these ruthless traders and stakers of claims for their routes and sometimes for their lives. Botanizing threatens no one. Botanists demand little. They are ideal explorers. And they can be exploited.

A dramatic opening for a great new century—and this small book about gardens—is to have the area suddenly doubled. To the modest established gardens of the settled East Coast of the so-recently-united states was suddenly added a vast wilderness. This, in an unforeseeably short time, would be expanded until it met in the West another ocean, in the South a great gulf, in the North a remote boundary drawn across lakes and forests. The seeds and scions of future gardens would travel in covered wagons or on horseback, later by steam and rail. They would be joined to and absorb other gardens in other styles—those of yesterday's enemies in Indian patches, formal French town gardens, enclosed Spanish patios. Log cabins among stumps in the developing inner countryside would become decorated cottages and pillared mansions.

With the Louisiana Purchase in 1803, the day of the early botanists and other followers of the infant sciences dawned. The world, as well as the United States, was waiting to see what they would find. The Old World was hungry for new plants, for strong trees, for fresh medicines, for more decorative shrubs and vines. And all the while specimens of everything of special interest were being collected. The trailblazers began to report and to mark down lands suitable for development and exploitation, even land to be avoided. They noticed and recorded everything of value. They made maps. Small wonder that several countries saw this expansion as an opportunity to send explorers of their own. A consideration of only a few of the early naturalists-explorers reveals what American gardens, and later those of the Old World, owe to them.

In the nineteenth century, the United States was a small world of people in a great world of wild nature. Characters, like the territories they explored, link, join, and impose themselves upon each other. No wonder the first quarter of the nineteenth century burst with new discoveries, influences, ideas, and systems, and with new people. New works were superimposed on the bases of old ones. Credits were sometimes generous, sometimes withheld, sometimes deliberately passed over. Personalities were as striking and dissimilar as their discoveries. And new gardens hung in every balance.

As in the eighteenth century, there were several "key" people and points of departure. M'Mahon's nursery and office, for instance, was a meeting place for all travelers in the Philadelphia area. A clearing house was needed. At the beginning of the century, Dr. Hosack in New York, Thomas Jefferson in Washington, and Dr. Benjamin Smith

Barton in Philadelphia were all senders-out. Later Asa Gray and John Torrey, self-designated "closet botanists," were taking in, occupied by, identifying, and recording what others had found. And of course there were also foreign visitors, obliged to make contact with them all before they could say they had seen the country.

There were established touch points for early botanical adventurers. Over and over again, they record arriving in New York, calling upon Dr. Hosack, proceeding to Philadelphia, and meeting the natural scientists there. Then, after running botanical errands north and south, they were drawn toward Pittsburgh and from there by boat or barge toward the West. Fortunately there was enough exploring to go around. Fortunately there were eager American naturalists and botanists established *in situ,* waiting to receive the adventurers and their specimens and to catalogue their finds.

Credits can never be fairly assigned to many of these enrichers of nineteenth-century American gardens. Their names recur in American garden flowers, as well as on our bookshelves. Their contributions are assessed by scientific societies with great herbariums and busy scholars. For simple gardeners, planting and cherishing their discoveries, the response is emotional and, of course, by their lights, abysmally ignorant. But we are grateful.

WHEN the area west of the original thirteen states was Spanish, Jefferson had been eager to know what it contained and had approached the Spanish ambassador with a suggestion that a small American scientific expedition be allowed to explore. The answer was instantly and unequivocally negative. But the idea remained with Jefferson, and with statesmen from other countries. Citizen Genêt, minister plenipotentiary of the new French Republic, saw the value of such a project and enlisted the distinguished French botanist André Michaux to undertake a journey through Spanish holdings in the interest of France.

André Michaux (1746–1802) arrived in New York in 1785 accompanied by his fifteen-year-old son, François-André (1770–1855), and a French gardener named Saulnier. French officials—in particular, le comte d'Angivillier, in charge of the king's parks and gardens—were alarmed by the depletion of French forests during the past centuries and were anxious to secure the services of an able botanist to explore in

"The Rocky Mountains: Emigrants Crossing the Plains" (Currier and Ives).
Courtesy the Collection of Heritage Plantation of Sandwich, Massachusetts.

America for trees to be used for French reforestation. Michaux had already carried off a successful exploration in Asia Minor at French government expense.[1] In 1785 he was about to be sent to Tibet, when,
happily for American horticultural records, his government decided to
send him to North America.

Since the beginnings of French travel anywhere, the French government had required subjects to observe, explore, and send back materials of horticultural interest. The Jardin du Roi in Paris was always ready
to receive their discoveries. Travelers were encouraged to bring back
seeds and plants and to set up small "holding gardens" to keep, test,
and grow their material until it could be transported to Paris. The
French Jesuit missionaries to the Indians in Canada were the first to
collect our native plants. Their ardor and efficiency explain why *canadensis* and *canadense* are so often a part of the botanical names of North
American wildflowers.

Upon his arrival in New York, André Michaux explored the area and settled upon a piece of land in New Jersey as ideal for his holding garden. He must have been naturally disarming, for he purchased the property from New Jersey officials despite a law prohibiting the sale of land to foreigners. Besides *les arbres utiles,* he had been directed to find suitable birds for French woodlands. At the end of his first year, he sent to France twelve boxes of seeds, five thousand seedling trees, and a shipment of live partridges.

By 1787 Michaux was in Charleston, South Carolina, where he decided to make his headquarters and a second garden. From there, he made extensive explorations—from Hudson's Bay to Florida's Indian River, from the Bahamas to the banks of the Mississippi. Spurred on by his success, he volunteered to explore beyond the Missouri if the American Philosophical Society would sponsor the journey, a subject that struck sympathetic chords in powerful places. Funds were raised—and subscribed to by, among others, Washington and Jefferson.

Michaux was able to extend his assignment farther west to visit French establishments; but his time in America was coming to an end, as was his financial allowance from the French government. When he found no fresh funds on his return to Charleston in 1795, he sailed for Paris but his ship was wrecked on the coast of Holland where he lost a large portion of his collections and papers. In Paris he discovered such ruin—due to neglect—of the once royal gardens where his plants had been grown that he determined to sail back to America to replace the loss. But the French government had also lost any interest and refused aid. Intead, Michaux had to accept a part in an expedition to New Holland in 1800; he died of fever in Madagascar in 1802. His great work on the oaks of America, *L'Histoire des Chênes d'Amérique,* was published in his absence in 1801, to be followed in 1804, after his death, by his *Flora Boreali-Americana.*

François-André Michaux, the son who had accompanied his father to America as a boy and learned his gardening from Saulnier, had been further instructed while in France by "des savants du Museum." He set out to try his own luck in the United States in 1801. He found the Charleston nursery safe, watched over by a neighboring planter, and the New Jersey nursery still cared for by Saulnier. The American Philosophical Society was still full of his father's friends. A recent French arrival in the United States, Du Pont de Nemours, tried to be helpful to his young compatriot. Like all foreign visitors, Michaux *fils* called upon Dr. Hosack in New York and on Dr. Mitchill, engaged upon his *Flora Carolinaensis*. After this brief reconnoiter, he returned to France and published an account of the naturalization of American trees, shrubs, and plants with such marked success that he was appointed *agent temporaire de l'administration des essences forestiers susceptibles d'être naturalisées* and in 1806 embarked again for the United States from Bordeaux.

On the high seas, Michaux's American vessel was stopped by a British frigate and he was taken to Bermuda. Here he was permitted to land and to make botanical excursions until he could sail for Philadelphia. From there, finally, and from Saulnier's little house in New York, he was able to explore the coast from Maine to Georgia and to journey to the Great Lakes. By 1807 twenty-five cases of his collections had arrived in France.

François-André Michaux returned for good to France in 1808 to finish his monumental *Histoire des Arbres Forestiers de l'Amérique Septentrionale*. The first of the three volumes appeared in 1810, the next two in 1813. Fortunately for us, he had engaged with John Vaughan, secretary of the American Philosophical Society, to have an English edition published, and the original manuscripts are still preserved in that society's archives. *North American Sylva* was instantly recognized as an important work. A second edition was destroyed by fire in 1857, but a third, with notes by S. S. Smith, appeared that same year in Philadelphia.

In retirement in 1822, Michaux built a house on property beside the Oise, where he also prepared his tomb. Surrounded by a grove of exotic trees, chiefly of American origin, he was buried there in 1855. In his will made in recognition of the "cordial hospitality" he and his father had received and to contribute to the advancement of agriculture and sylviculture in the United States, he left a handsome sum to each of the two societies of which he had been a member: the Society of Agriculture and the Arts in Boston and the American Philosophical Society of Philadelphia.

THE time had come for Americans to explore for themselves. Meriwether Lewis (1774–1809) seemed groomed by fate to find a way to the Pacific across the newly acquired Louisiana Purchase and on over expanses of uncertain and disputed ownership. He was, in fact, carefully prepared by Thomas Jefferson to undertake a task in which Jefferson himself had a longstanding special interest.

Born on a plantation near Charlottesville, Virginia, Meriwether Lewis explored the countryside as a boy until his father died and his mother remarried and moved to Georgia. Lewis returned to Virginia at the age of thirteen to be tutored for William and Mary College, but when he was seventeen his stepfather died and his mother returned to Locust Hill, near Jefferson. Lewis settled in to manage the estate for her. As Jefferson wrote of him later, he was acquiring "a talent for observation which had led him to an accurate knowledge of the plants and animals of his own country." Lewis was considered too young, at eighteen, for the Michaux expedition, but at twenty he volunteered when there was a call for troops to suppress the Whiskey Rebellion. From across the mountains near Pittsburgh, he reported to his mother that, "delighted with a soldier's life," he had enlisted in the regular army. Commissioned an ensign, he marched to Greenville, Ohio, where the treaty made by Anthony Wayne with the northern Indians brought peace to the Northwest Territory. During this campaign, Lewis served under William Clark (1770–1838), whom he was later to select as his companion in exploring the West. Lewis saw active duty at Fort Pickering in Chickasaw Indian Territory and learned the language and customs of those Indians.

Lewis was at the outpost of Detroit when Jefferson, newly elected president, wrote (during his first week in office) to offer Lewis the post of his private secretary. Jefferson said he knew of Lewis's "capacity to aid in the private concerns of the household," but he valued also the mass of information that Lewis had acquired about the western regions. Soon after the inauguration, Lewis moved into the White House to take charge of all domestic arrangements for the widowed Jefferson. From the conversations of congressmen, diplomats, and friends like Thomas Paine, young Lewis could not fail to learn. Jefferson sent Lewis to read his annual message to the Senate on December 8, 1801.

Ever since 1792 when Lewis, at the age of eighteen, had begged Jefferson to let him help explore for a land route to the Pacific Ocean, the subject had never been abandoned. In the winter of 1803 Jefferson asked Congress for an appropriation for a journey to study the far western Indians. Congress granted the $2,500 that Lewis estimated would be necessary.

In 1814, writing his account of the Lewis and Clark expedition, Jefferson unstintingly praised Lewis's qualifications:

> courage . . . perseverance . . . careful as a father of those committed to his charge . . . steady in the maintenance of order and discipline . . . intimate with Indian characters . . . habituated to the hunting life . . . guarded by exact observation of the vegetables and animals of his own country against losing time in the description of objects already possessed . . . a fidelity to truth so scrupulous that whatever he should report would be as certain as seen by ourselves.

Before the expedition started, however, though it seemed to Jefferson that Lewis's qualifications had been "implanted by nature in one body for this express purpose," Jefferson undertook to have what nature had not yet effected made up for in Philadelphia. He sent Lewis there to be briefly instructed in astronomy, mapmaking, botany, and the collecting of material. Jefferson himself had made a long list of instructions, which he sent to Lewis in 1803, with passports through French territory (suddenly rendered unnecessary when the Louisiana Purchase was confirmed). Part of Jefferson's instructions included a record of each day's discoveries on paper in at least three copies. An added precaution was to make an additional copy on birch bark, which would be impervious to water damage. The last touch proved to be the most practical. After a shipment was made from Fort Mandan, all the rest of the material was lost at Great Falls. A collection made on the way back was delivered by Lewis himself to Jefferson, to be stored until the two explorers returned to sort and record their findings.

Jefferson had arranged for Lewis to choose a companion officer, and William Clark of Louisville met Lewis at St. Charles, Missouri, in the spring of 1804. By fall, the expedition had reached the Mandan villages in North Dakota and was not to be heard from again for eighteen months.

The collection and identification of plant materials by Lewis and Clark was awaited with great eagerness by botanists and all "curious" American gardeners. Jefferson sent the shipment "of plants, earths and minerals" received from Fort Mandan to the American Philosophical Society, with a request that it be delivered to William Hamilton, proprietor of Woodlands near Philadelphia. Hamilton was one of our leading horticulturists, and the material was to be planted and reported on by him. The dried specimens were to go to Dr. B. S. Barton for identification. The seedsman Bernard M'Mahon was also alerted and was also sent seeds.

THE interest in the new plant discoveries was intense and elaborate precautions were taken for their safety. When much later, on Lewis's return, M'Mahon learned that Lewis himself had brought back a second lot of dried, preserved plants, he wrote to Lewis in great excitement. As Lewis was said to be coming to Philadelphia, M'Mahon requested that he hurry to be there before a "young man boarding at my house who in my opinion is better acquainted with plants in general than any man I ever saw" could leave town.

This young man, Frederick Pursh, a botanist, had been "bred to the business in Saxony" and had lived with William Hamilton ("who between you and me did not use him well") for two years. For the past year, he had been employed collecting for Dr. Barton. Pursh was working for Barton, living with M'Mahon, and harboring some sort of grudge against Hamilton. There are no descriptions that would endear him to us. Pursh was about to depart on another mission for Barton but agreed to delay to meet Lewis. M'Mahon could not have foreseen the horticultural complications that would ensue from this meeting.

Lewis met with Pursh in early May of 1805 and paid him thirty dollars, with an additional forty "for assisting me in preparing drawings and arranging specimens of plants for my work." At the end of May, in all good faith, Lewis turned over to Pursh his entire collection of herbarium specimens. Pursh wrote that Lewis had wished him to "describe and figure those I thought new for the purpose of inserting them in the account of his *Travels* which he was then engaged in preparing for the press."[2] In his introduction to his great work, *Flora Americae Septentrionalis,* Pursh said the collection, of about 150 specimens, contained not above a dozen plants well-known to him.

Clark arrived in Philadelphia in 1810 to locate Lewis's materials and to hasten the publication of the Lewis and Clark *Journals*. He sent to New York for Pursh's sketches and reimbursed Pursh for what Pursh said were "all the drawings prepared for the work." These, and what Clark believed to be the complete collection entrusted to M'Mahon, he had the latter deliver to the ailing Dr. Barton, who had again agreed to undertake the scientific accounting of the expedition. Barton's ill health again prevented him from realizing what would have been a crowning achievement for both himself and his country. As Dr. Barton's failing health had prevented him from being able to describe Lewis's collection after he had received the first shipment from Fort Mandan, Pursh became the only one to receive the rest of the specimens from Lewis on his return.

By 1807 Pursh had completed his work for Barton, Lewis had gone to take up his new appointment as governor of Louisiana, and M'Mahon had become uneasy with Pursh working in his house on Lewis's material. Although Lewis had deposited it there for Pursh to study, M'Mahon still felt responsible and he wrote Jefferson that he was reluctant to let Pursh leave until Lewis could meet with him. Only Lewis's "particular instructions" could permit Pursh to "finish the drawings of some very important specimens." When he left Louisiana for Washington in the fall of 1809, Lewis must have been planning to complete this work. We shall never know, as he committed suicide or was murdered on the Natchez Trace on October 11.

Pursh had left M'Mahon's to go to New York to work with Dr. Hosack. M'Mahon wrote Jefferson again when he heard of Lewis's death, saying that he believed that he had all of Lewis's original collection of dried plants but that "Mr. Pursh had taken his drawings and descriptions with him." M'Mahon warned that Pursh would "no doubt on delivery of them expect a reasonable compensation for his trouble." M'Mahon also asked Jefferson's advice on another matter. He himself had grown several kinds of "*new* living plants," which he had raised from seed, but he was reluctant to part with these into the hands of those "who would gladly describe them and publish them without doing due honor to the memory and merit of the worthy discoverer."

Anxiously, M'Mahon added he was sure Jefferson was interested in having "the discoveries of Mr. Lewis published" and so felt it incum-

bent upon himself to give the preceding information and to ask advice
as to the propriety of keeping the living plants. What M'Mahon did
not know then was that Pursh had taken with him to New York not
only all the work he had done but also whatever of Lewis's pressed
plants he had pleased to select, both duplicates and—in the case of
simple specimens—either part of the plant or the whole. It is hard to
understand Pursh's plan of procedure because he continued to work
with Dr. Hosack until the winter of 1811–12, when he left for England. He began work then on his *Flora Americae Septentrionalis,* which
contained 124 plants collected by Lewis and Clark, designated "V. S.
in Herb Lewis." Of the twenty-seven illustrations, thirteen were from
Lewis's herbarium. In a last-minute flourish of honor, Pursh created
two genera named *Lewisia* and *Clarkia.* Pursh died in 1820. To Pursh
alone must now go the credit for saving, however dubiously, the collection so laboriously made by Lewis and Clark.

LIKE reiterations and confirmations of those initial quests, we shall see
the same procedures repeating themselves. The case of John Charles
Fremont illustrates this principle. Headed west in a series of explorations aimed at extending the sovereignty of the United States to the
Pacific shores, his adventures make an extraordinary tapestry of con-
flicting patterns, woven across trails of Indians and trappers and settlers,
to include finally the eventual crossing of the continent by railroad.

Fremont's first appointment in 1823 was by President Jackson, who
had been authorized by Congress

> to employ two or more skilful civil engineers . . . to cause the
> necessary surveys, plans and estimates to be made of the routes of
> such roads and canals as he may deem of natural importance in a
> commercial or military point of view or for the transportation of
> the public mail.

The first of Fremont's five expeditions occurred in 1842, the last in
1854, when he established to his own satisfaction "the favorable nature
of the central route for a railroad in winter as well as summer."

While he mapped and sketched and fought and planned, he collected
flowers. Some have been lost, as always, but many—300 species—have
been accounted for by our friends Asa Gray and John Torrey. Gray, im-

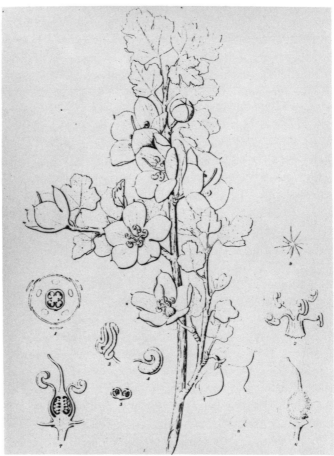

Fremontia californica (from Smithsonian Institution).

patient for publication of the collection, published a memoir entitled *Descriptions of Some New Plants Collected by J. C. Fremont in California,* illustrated by Isaac Sprague of Cambridge, Massachusetts. This was "accepted for publication" in 1850 by the Smithsonian *Contributions to Knowledge.*[3]

NOTES TO CHAPTER 2

1   After two years, he was rescued from this voyage and its extreme hardship by the British consul general at Basra, and returned to Paris loaded with seeds.

2   In the introduction to his *Flora Americae Septentrionalis.*

3   Two of the illustrations commemorate *Fremontia californica* (see illustration in this chapter) and *Spraguea umbellata,* an interesting example of the obscure credits nestled in botanical nomenclature.

THREE

Those Blessed Visitors: Observers

from Abroad

ROBERT BURNS to the contrary, the gift of seeing themselves as others saw them in the early nineteenth century was a free gift to Americans, if one can except what it cost to act as host to the critics—to host the host, for critics came in droves. The nineteenth century, as we know, was a touchy time everywhere. The old orders were changing as always, the common man was emerging and flexing his muscles—especially those used in gardening. Glass-housing of plants was being developed in England on a great range of social scales. Social travel—that is, for pleasure—was beginning to be common, and trips even so far as the next county became more frequent.

To go as far away from home as possible, especially across the Atlantic, made one an English celebrity, increasing the sales of one's books and lectures even on quite different subjects and insuring fortunes to be made for writing and speaking after one returned "home." As was well known, the new United States had little to offer in the way of cultural uplift in the arts—painting and music and sculpture requiring a certain amount of security, leisure, and access to the proper materials—but it had a new form of government (no matter how certain many felt that it was to fail), and it had ideas about social reform. From abroad, men of letters came gravely to judge, bringing their wives, snippily to write. Ladies with the minds of men, given to writing and speaking on their own, arrived to visit interminably. A stay of two years was considered respectable, especially if the journey included Canada.

The first great shock for visitors coming into East Coast parts was, after the wildness of the landscapes, the condition of the streets of American cities—pure sand. And then there was the state of American roads—pure mud.

Visitors in the early nineteenth century fell into two categories. The first was made up of those who were impressed by the idea of a whole new form of government and wished to see how—and if—it worked. Those in the second category came to make paying audiences of those they visited, to coin money from their visits, and to write books about their harrowing social experiences and the conditions of American roads, coaches, and hotels. These are the ones read today, no matter how annoying their observations.

For our purposes, sadly enough, the noble visitors were not generally given to describing gardens, although Harriet Martineau's description of the way Canadian Indians were treated in 1834 includes a few flowers.

On her way to Mackinac by steamer, she observed that

the island looked enchanting . . . steeped in the most golden sunshine that ever hallowed lake or shore. . . . The better houses stand on the first of three terraces. . . . Behind them are swelling green knolls; before them gardens sloping down to the narrow slip of white beach, so that the grass seems to grow almost into the clear rippling waves. The gardens were rich with mountain ash, roses, stocks, currant bushes, springing corn and a great variety of kitchen vegetables.

Frances Trollope arrived in the United States in 1827. She had come intending to make money not from writing but from opening "an emporium" in one of the frontier settlements, Cincinnati. It was her husband's idea. Middle-aged but still eager to try new ideas in new places, Mrs. Trollope arrived at the mouth of the Mississippi with three small children at Christmas time. The scene, she felt, surpassed in desolation anything described by Dante. But that was only the beginning. The two days to New Orleans seemed longer, in their "total want of beauty," than the entire ocean voyage. On landing, the heat was "much more than agreeable and the attacks of mosquitoes incessant," but the sight of oranges, green peas, and red peppers growing in the open was cheering. Mrs. Trollope had resolved not to allow "the fidelity of description" to escape her. Even when they reached Memphis, bound by steamer for Cincinnati, "at the most beautiful point of the Mississippi" she found the dark forests at either end of the cleared space oppressive.

Leaving Memphis after five days, the Trollopes set out for Cincinnati, which they had been assured was the "finest situation west of the Alleghenies" and rejoiced to leave the muddy Mississippi for the clear and bright Ohio. The primeval forest still lurked, but there was

almost every variety of river scenery . . . a meadow of level turf . . . perpendicular rocks . . . pretty dwellings with their gay porticoes. . . . often a mountain torrent comes pouring its silver tribute to the stream, and, were there occasionally a ruined abby or a feudal castle to mix the romance of real life with that of nature, Ohio would be perfect. . . . yet these fair shores are still unhealthy.

Time only confirmed her aversion to Ohio, where she found the hills afforded "neither shrubs nor flowers." The Trollopes finally left; in three days by steamer they reached Wheeling, West Virginia. On they went to Baltimore, where Mrs. Trollope seems to have taken heart, finding it a "beautiful city." Her outlook and even her tune changed somewhat, at least with regard to gardens. Mrs. Trollope's *Domestic Manners of the Americans* reported on what she saw. Its editor, John Claudius Loudon, proclaims that we can trust her observations because she was known not to like Americans. In 1832 she reported on a variety of American scenes, the most horticulturally oriented of which are excerpted here, from an account compiled by Loudon from Mrs. Trollope's descriptions.

Hyde Park is the magnificent seat of Dr. Hosack; here the misty summit of the distant Kaatskill begins to form the outlines of the landscape; and it is hardly possible to imagine a more beautiful place.

The neighbourhood of Philadelphia is rendered interesting by a suc-cession of gentlemen's seats on the Delaware, which "if less elabo-rately finished in architecture and garden grounds than the lovely villas on the Thames, are still beautiful objects to gaze upon as you float rapidly past, on the broad silvery stream that washes their lawns. . . ."

Stonington is about two miles from the most romantic point on the Potomac River; and Virginia spreads her wild but beautiful and most fertile paradise on the opposite shore. The Maryland side partakes of the same character, and displays an astonishing profusion of wild fruits and flowers. The walk from Stonington to the falls of the Potomac is through scenery that can hardly be called forest, park, or garden; but which partakes of all three. Cedars, tulip trees, planes, sumachs, junipers and oaks of various kinds, shade the patch. Below are Judas trees, dogwood, azaleas, and wild roses; while wild vines (*Vitis vulpina?*) with their rich expansive leaves and sweet blossoms rivalling the mignonette in fragrance, cluster round the branches; and strawberries, violets, anemones, heartsease, and wild pinks literally cover the ground. The sound of the falls is heard at Stonington, and the gradual in-crease of this sound is one of the agreeable features of this deli-cious walk. A rumbling, turbid, angry little rivulet, called the Branch Creek, flows through evergreens and flowering under-wood, and is crossed *a plusieras reprises* by logs thrown from rock to rock. The thundering noise of the still unseen falls suggests an idea of danger while crossing these rude bridges, which hardly belongs to them; and, having reached the other side of the creek, the walk continues, under the shelter of evergreens, another quar-ter of a mile, and then emerges on the rocky depths of an enor-mous river; and so large are the black crags that enclose it, that the thundering torrents of water rushing through, over, and among the rocks of this awful chasm, appear lost and swallowed up in it.

The whole region of the Allegheny mountains is a garden. The mag-nificent rhododendron fringes every cliff, nestles beneath every rock, and blooms around every tree. The azalea, the sumach, and every variety of the beautiful mischief the kalmia, are in equal profusion. Cedars, first, and the hemlock spruce attain here the greatest "splendour and perfection of growth." Oak and beech, with innumerable roses and wild vines hanging in beautiful confusion among their branches, were in many places scattered among the evergreens, and the earth was carpeted with various mosses, and creeping plants. Often, on descending into the narrow valleys, spots were found in a state of cultivation. These little gardens, or fields, were hedged around with sumachs, rhododendrons, and azaleas; and the cottages were covered with roses. The valleys are spots of great beauty, and a clear stream is always found running through them, which is generally converted to the use of the miller.

At New York . . . The park, in which stands the noble city-hall, is a very fine area. . . . There are a few trees in different parts of the city, and many young ones have been planted, and guarded with much care. . . . The enclosure in the center of Hudson's Square (New York) is beautiful. It is excellently well planted with a great variety of trees, and only wants our frequent and careful mowing to make it equal to any square in London. The iron railing which surrounds this enclosure is as high, and as handsome, as that of the Tuilleries; and it will give some idea of the care bestowed on its decoration, to know that the gravel for the walks was conveyed by barges from Boston, not as ballast, but as freight.

Hoboken, on the North River, about three miles from New York, is a public walk of great beauty. . . . A broad belt of light underwood and flowering shrubs, studded at intervals with lofty forest trees, runs for two miles along a cliff. . . .

In the city of Washington there are several squares newly planted; and some of the streets are bordered by rows of trees. The avenue of Pennsylvania, when the trees are a few years older, will be the finest street in the world. . . .

At Baltimore the public walk is along a fine terrace . . . ornamented with a profusion of evergreens and wild roses.

Waterworks at Fair Mount, near Philadelphia. Fair Mount is one of the prettiest spots the eye can look upon. . . . On the farther side of the river is a gentleman's seat, the beautiful lawn of which slopes down to the water's edge; and groups of weeping willows

"The Village of Riceborough is very picturesque."

"Occasional villages gave some relief to the tedium of this part of the
journey through the pine barrens of Georgia."

Two Basil Hall sketches, taken from J. C. Loudon's "History of Gardening" section, in his *Encyclopaedia of Gardening*, edited by
Mrs. Loudon, 1871.

and other trees throw their shadows on the stream. The works
themselves are . . . in a simple but handsome building . . . [be-

hind which] rises a lofty wall of solid limestone rock, which has at one or two points been cut into, for the passage of water into a magnificent reservoir. . . . From the crevices of this rock the catalpa was every where pushing forth, covered with its beautiful blossoms. Beneath one of these trees, an artificial opening in the rock gives passage to a stream of water . . . received in a stone basin of simple workmanship, having a cup for the service of the thirsty traveler. At another point, a portion of the water . . . is permitted to spring forth in a perpetual jet d'eau that returns in a silver shower upon the head of a marble naiad of showy whiteness, admirably relieved by its dark rocky background, and the flowery catalpas which shadow it.

Promenade at Philadelphia. There is a very pretty enclosure before the walnut-tree entrance to the state-house, with good well kept gravel walks, and many beautiful flowering trees. Near this enclosure is another . . . called Washington Square, which has numerous trees, with commodious seats placed beneath their shade.

Mrs. Trollope even offers a kindly recollection of the city she so longed to leave.

At Cincinnati there is a public garden, where the people go to eat ices and look at roses. For the preservation of the flowers, there is placed at the end of one of the walks a kind of signpost, representing a Swiss peasant girl holding in her hand a scroll, requesting that the roses might not be gathered.

For his own *Encyclopaedia of Gardening,* Loudon was fortunate to be able to draw upon Mrs. Trollope for his direct quotations upon the subject of gardens in North America. Three illustrations are from the pen of Captain Basil Hall: a small house in Riceborough, Georgia; a pine forest; and a log-cabin clearing in the wilderness. The only two other illustrations are one of a "worm fence" and Downing's sketch of the Lyman estate near Boston.

Free Spirits and Fixed Points

N addition to the official botanists of the early nineteenth century, sent by governments in search of all sorts of knowledge, there were individual enthusiasts for botanical exploration, willing to go anywhere and bear with anything to discover new plants. These collectors were the really free spirits. In some cases, their purpose was to explore in limited but well-defined areas near their homes and to report to local authorities and societies. Sometimes they would join with one another and make expeditions to nearby mountains. Early, self-trained botanists like the Bartrams, father and son, who collected plants for the love of them, were willing to be subsidized by wealthy Englishmen, far across the Atlantic, who awaited their "boxes" with impatience, gentlemanly greed, and requests for further trips into ever more remote wildernesses. Because of the success of Peter Collinson, the Quaker haberdasher and naturalist, in introducing the Bartrams' American plants to private English gardens, English societies and nurseries began to send out their own emissaries.

By the nineteenth century, English horticultural and botanical societies were setting their sights on lands as remote as Asia and South America. Various arboretums and pinetums coaxed adventurers to explore for them on the American West Coast, where trees were reported to be gigantic beyond imagining. Modest nurseries and groups of

wealthy friends became anxious to be the first to introduce new flowers
to Victorian garden schemes.

A quotation from an article, "On the Botany of America," written
in 1825 by William J. Hooker of Glasgow, the first to preside over the
Royal Horticultural Society, puts clearly the desirability of importing
American plant material.

> What renders the botany of North America peculiarly interesting
> to the British Naturalist is that a very large proportion of its
> vegetable products may be assimilated to our own climate. The
> oaks and firs now decorate many of our plantations and pleasure
> grounds. . . . Our shrubberies owe their greatest beauty to the
> various species of Kalmia, Azalea, rhododendron, robinia, cor-
> nus, sambucus, ceanothus, lonicera, syringa, flowering rasp-
> berry . . . which flourish as if they were aboriginal natives of our
> soil . . . our most highly flavored dessert apples are imported
> from America . . . we now procure grafts for orchards and wall
> trees . . . and for the curious we have Dioneae and Sarracenia.

By the nineteenth century, the search for new material for English gardens had become a social rage. Expeditions to Mexico, Peru, California, and Texas had discovered a wealth of brightly colored plants,
which incredibly enriched English herbaceous borders with hardy perennials and contributed to carpet and ribbon bedding with gay little
plants that had to be started under glass and treated as annuals. Magazines regularly celebrated new finds with scientific details and beautiful illustrations. Even today, the uplands of Texas wave for miles with
the knee-high blossoms of lupines, phlox, coreopsis, and gaillardia,
just as they did when first seen by lonely botanists.

Some of the adventurous plant explorers were not above collecting
for pay and in bulk for English purchasers. One of these, Thomas
Drummond, explored first under the aegis of William Hooker but felt
free to collect Texas wildflowers in quantities that would allow distri-
bution to subscribers. In a short time, in the then much troubled
Texas, Drummond survived cholera and floods to send back an incredible number of new plants—700, it was claimed, and a quarter that
many birds. He collected generously of all the more interesting plants
for his patrons, such as Hooker and the newly formed Horticultural

Society in London. On account of his industry, many plants were named after him by those fixed stars of botanists who worked in their laboratories to identify and publish the finds of those in the field, the fixed points in this universe. Little is known today about Thomas Drummond, except that he spent a few years among the Texas flowers that carry his name into our gardens every summer. And who would not be happy to be immortalized by our border stalwart, *Phlox drummondii?*

There were also individuals who botanized as they pursued other careers, like Joel Poinsett of Charleston, South Carolina, who was one of the best traveled and most sophisticated Americans of his time. Sent on secret missions to South America and to Mexico by Presidents Madison and Monroe, he was made first American minister to Mexico by President Jackson and was appointed Van Buren's secretary of war before he finally retired to his Charleston gardens, full of the plants he had collected from around the world. The plant that carries his name, *Poinsettia pulcherrima,* was exhibited in Philadelphia at the "First semi-annual Exhibition of fruits, flowers and plants" in June 1829, where mention was made in the reports of the "new Euphorbia, with bright scarlet bracteas, or floral leaves, presented to the Bartram collection by Mr. Poinsett, United States Minister to Mexico."

AND of course there were the total eccentrics, unable ever to stop botanizing, such as the extraordinary Constantine Rafinesque. He has left his own account of his joys and trials in the prologue to his *New Flora of North America,* published in 1836.

> I always traveled with my botanical collecting book and reams of paper to preserve my plants. . . . I have been able to collect in 20 years . . . a most valuable Herbarium, rich in new species, rare plants and complete Monographs. . . .
>
> During so many years of active and arduous explorations I have met, of course, all kinds of adventures, fares and treatment. . . .
>
> Let the practical Botanist who wishes like myself to be a pioneer of science and to increase the knowledge of plants, be fully prepared to meet dangers of all sorts in the wild groves and mountains of America. The mere fatigue of a pedestrian journey is nothing compared to the gloom of solitary forests where not

a human being is met for many miles, and if met he may be distrusted. . . .

You meet rough or muddy roads to vex you and blind paths to perplex you, rocks, mountains, and steep ascents. You may often lose your way and must always have a compass with you. . . . You may be lamed in climbing rocks for plants or break your limbs. . . . You must cross and wade, . . . in deep streams you may lose your footing. . . . You may be overtaken by a storm, the trees fall around you. . . . The fire of heaven or of men sets fire to the grass or forest and you may be surrounded by it . . . over an unhealthy region or in sickly season you may fall sick on the road.

On the other hand,

I never was healthier and happier than when I encountered those dangers. . . . I like the free range of the woods and glades. I hate the sight of fences like the Indians! . . . The pleasures of a botanical exploration fully compensate for miseries and dangers . . . fair days and fair roads . . . a clear sky . . . the pure air of the country . . . what soothing naps at noon under a shaded tree near a purling brook!

Every step taken . . . appears to afford new enjoyments . . . you feel an exultation . . . you are going to add a new object or a page to science. . . . Such are the delightful feelings of a real Botanist, who travels not for lucre or filthy pay. . . . When nothing new nor rare appears, you commune with your mind and your God in lofty thoughts or dreams of happiness. Every pure botanist is a good man, and a religious man. . . . To these botanical pleasures may be added the anticipation of the future names, places, uses, history etc. of the plants you discover.

Rafinesque—eccentric, genius, and budding naturalist—had arrived in Philadelphia from France in 1802, before he was twenty, to work as a clerk in a countinghouse. On the side, he worked at classifying the birds in Peale's Museum (no small task for a newcomer) and botanized so enthusiastically and successfully in the Philadelphia region that he decided to write the first comprehensive *Flora* of the Middle Atlantic states.[1] Heady with the possibilities of the New World, Rafinesque purposefully returned to the Old for ten years, which he spent in Sicily, trading in medicinal plants and writing papers for Dr. Mitchill's *Medical Repository*. He also wrote, among other papers, a study of European plants that had become naturalized in America. He studied ichthyology as well, following Linnaeus's classifications. By 1815 he had amassed several collections (including one of shells), a large library, and a stock of botanical drugs.

Having made this substantial investment in his future, Rafinesque happily set sail for the United States, only to be wrecked off Block Island after a disastrous voyage. He arrived destitute at the home of Dr. Mitchill. But his spirit had not been quenched; this continent still held a wealth of undiscovered material. In New York he met the young John Torrey, already a fixed star, with whom he could discuss his preference for the natural system of Jussieu over the sexual system of Linnaeus, which was so loyally espoused by most Americans. He went off botanizing down the Ohio toward Kentucky, which he later was to make peculiarly his own territory. The new University of Transylvania in Lexington asked him to teach natural history, and he performed with such enthusiasm and originality, using living botanical specimens to illustrate his lectures, that no one who had sat in any of his classes ever forgot him. (Indeed, Jefferson Davis, one of Rafinesque's pupils, sent for his old textbooks on botany and conchology to while away time in prison after his capture.)

It is sad to realize that Rafinesque's enthusiasms betrayed a vulnerability that was exploited by contemporaries. Audubon, for example, amused himself by describing fish that had never existed and left his gullible and loyal guest to be disgraced for publishing them in good faith.[2] Rafinesque's real successes, such as being the first man in America to receive a gold medal from the Geographical Society in Paris and being designated by Swiss scientists "Catesbaeus" (as a worthy successor to the first author of a study of American birds), may have goaded his critics. And they were many.

Modern critics feel his greatest failure was a lack of organization. Present-day researchers find he could have saved followers many hours by keeping better records of his—it now appears—extensive discoveries. Today his credits mount.

Dismissed from the university in 1826, he traveled on the new Erie Canal, "One of the most agreeable journeys I ever performed," and reached Philadelphia still full of projects and enthusiasm. He published a modest *Medical Flora,* which must have seemed to bring him almost full circle, and wrote an enormous poem on the whole world, which he did not sign, as he intended to review it himself. Sadly, he died a pauper in Philadelphia, where only the intervention of friends—come a little late—saved his body from being sold to medical students by his landlord.

His hymn to the joy of collecting must be allowed to stand at the head of descriptions of his life's enthusiasms. A plant that bears his name today carries a certain poignancy. *Viola rafinesquii* is, according to L. H. Bailey, "indigenous to the U.S. from N.H. southward and westward to Texas and Colorado" and is a "delicate species" of the European wild pansy, heartsease, or Johnny-jump-up.

ONE of the most dedicated of the early nineteenth-century plant hunters in America was Thomas Nuttall. Born to a hard-working Yorkshire family, he was apprenticed at the age of fourteen to an uncle who was a printer in Liverpool. Realizing that his inclinations were likely to demand more education than he had been able to obtain at home, Nuttall educated himself in Liverpool in his spare time, in Latin, Greek, and French, and so outran his strength that he was sent home to rest and improve his health. Like other great men who in their youth found solace and future careers while wandering the countryside, Thomas took to the hills and fields with another youth of his age who was already deep in the study of botany. With John Windsor, whom he would see again in Philadelphia, he climbed in the Pennines, collected plants, and began to be interested in geology. The relationship between plants and the soil on which they grew added mineralogy to his scientific pursuits. By the time he returned, restored in health, to his Uncle Thomas, he had decided to study natural history, to travel (if possible, to the United States), and to rely upon printing only as a means of support. An angry uncle let him nearly starve during a year in London, but Thomas managed to sail for Philadelphia in 1808.

Thomas Nuttall began to botanize as soon as his feet touched American soil, and within a day he had questions to ask about plants he had found. He needed an American botany (one wonders if he then determined to write one some day) and applied to Benjamin Smith Barton himself when he failed to find a copy of Barton's *Elements of Botany.*

Fortunately for Nuttall, Pursh had just finished collecting for Barton and had departed from Philadelphia. Into his place went the young Nuttall, to use Barton's library and to be instructed by him.

Anyone reading about the United States in the nineteenth century will soon realize that Philadelphia was the center of scientific inquiries of all sorts. For many years it remained the place for keeping posted on all expeditions that pertained to the study of natural history. Another evident fact is that, although the New World was vast, the world of interesting and interested people was intimate. And kind. Repeatedly, we see strangers applying to busy scholars and physicians and authors and collectors and gardeners in Philadelphia and being received, even welcomed, into that industrious fraternity.

Nuttall's association with Barton led to his making the acquaintance of the elderly William Bartram and to his introduction to Peale's Museum. This last was an extraordinary showcase of exhibitions of many interesting items collected throughout the new continent. Charles Wilson Peale, a distinguished portrait painter and a keen amateur naturalist (with sons named Rembrandt, Rubens, and Titian, similarly inclined), had an urge to unite all phases of education and enlightenment in the arts and sciences. Somewhat like the Tradescants in the late seventeenth century, who had collected intriguing bits and pieces from all over the then known world to form their "Ark" (which later became the foundation for the Ashmolean Museum), Peale's Museum in Philadelphia covered an incredibly wide range of interests.

Scientifically arranged, according to the Linnaean system, were what might be called curiosities, embracing all things animal, vegetable, and mineral. Under the portraits of prominent Americans stood specimens of animals, birds, and fishes. Some of the specimens were arranged in habitat groups, an idea initiated by Peale. Insects were represented in quantity, as were fossils, reptiles, and shells. There was even a mastodon skeleton, which Peale himself had dug up from the Hudson valley, and there were relics of the Indians, brought back by Lewis and Clark and deposited in the museum at the suggestion of Thomas Jefferson. For young enthusiasts of the developing natural sciences, Peale's Museum must have acted as a catalyst, firming up for them the priorities of their life interests and offering them new ranges of inquiry.

Nuttall gravitated naturally to the nursery offices of Bernard M'Mahon, whose *American Gardener's Calendar* had been published in 1806 with great success. M'Mahon's offices served as meeting ground for a wide group of the area's keenest gardeners—those who aspired to the collection and growing of rare plants—and of scientists eager to obtain new material.

It could have seemed a perfect world to a young man longing to set out on his life's adventure of collecting and identifying plants, birds, fossils, rocks—and of studying the habits of primitive people. In every category, there was someone willing to help, to promote, to receive specimens, to assist in identification. Pursh had departed only recently; the rifts of jealousy and competition, the accusations of plagiarism— and the pranks to trip the unwary—were far ahead.

Nuttall must have thought he had found the ideal base from which to launch himself as soon as he could afford to leave his trade of printing. To Dr. Benjamin Smith Barton, Nuttall's arrival must have appeared equally providential. Ever anxious to write the definitive book on American flora, ever unable (due to the pressures of medical businesses and his own failing health) to undertake his own explorations, he had to depend upon the young and vigorous, even upon the "innocent," to collect for him. In all fairness, Barton may have been one of the innocents himself, in that he did not know what he asked when he sent his protégés into the unexplored wilderness.[3] Barton had distant and daring goals in mind for this new young man, recently from England, eager to find new plants and birds—but unable to swim, to shoot, to make fires, and (most important of all) to worry about his own fate when confronted by a wealth of plants and birds never before seen or recorded.

After sending Nuttall on a short trip to the South in 1809, as a test of his possible usefulness, Barton, anxious to further his account of the Lewis and Clark expedition, saw a way to advance his literary ambitions. For eight dollars a month and all expenses, Nuttall was to travel, keeping a journal for Barton's use alone and collecting "animals, vegetables, minerals, Indian curiosities, etc.," also for Barton (although Nuttall could keep some strictly for himself, to prevent their falling into the hands of others who might use them to Barton's disadvantage—a provision on which future misunderstandings would catch).

A contract was signed. When Nuttall later met another Englishman in St. Louis, John Bradbury, who had been sent out to collect plants for the Liverpool Botanic Garden—for three years, at a hundred pounds a year, with an additional commission for collecting bird skins—Nuttall may have entertained misgivings about his own bargain. But he appears not to have harbored grudges.

The original idea had been for him to proceed across the country, as if there were little to impede his progress overland, until he reached a Northwest barely touched by Lewis and Clark. Barton would have a greenhouse built for raising the seeds Nuttall would collect and a garden made for the plants. There were to be no slips between cup and lip, as there had been with the Lewis and Clark expedition's seeds and plants. Equipped with a gun and powderhorn, and with materials for making his collections, Nuttall set off for Pittsburgh in a mail coach, early on a spring morning in 1810.

After eight days in the coach, he arrived in Pittsburgh, where he spent a few days and found his white dog tooth violet (*Erythronium albidum* Nutt.). Then he started on foot for Lake Erie. He was frequently stricken with fits of malaria but was rewarded by the discovery of four species which now bear his name, of which *Collinsia verna* Nutt., named for the Philadelphia philanthropist, became a lifetime favorite; it is described as having a lower lip of "bright azure blue which no color can excel." From Erie, he set out to walk to Detroit—all boat arrangements having failed—but at Detroit he realized that Barton's suggested land routes were probably unrealistic. He took to the more reliable waterways. In a canoe. With Indians.

Bedazzled as we are, and should be, by Nuttall's tenacity and unrelaxing enthusiasm for collecting in the wilds, we gardeners cannot here follow him through the details of this, his first major adventure.[4] We can only note that after he finally arrived in New Orleans, miraculously alive and with all his collections for Barton shipped safely to Philadelphia, Nuttall seized an opportunity to sail to England and stayed there for the war years.

He did not return to Philadelphia until 1815, by which time all the collectors' ferments were bubbling. Barton had left to visit England; Muhlenberg and William Hamilton had both died. Zaccheus Collins still befriended all and sundry, and Abbé Correa da Serra taught the

Jussieu natural system in Barton's botany courses. Public lectures absorbed the scholarship and attention of many. Jacob Bigelow was reported to be giving his popular lectures on botany in Boston to help the Cambridge Botanic Garden (of which, incidentally at this point, Nuttall would later be in charge).

In Philadelphia, Nuttall had some ideas of his own. He had seen Pursh's *Flora* in England and longed to correct and add to it. M'Mahon was sympathetic and interested. The book had not yet reached America. Perhaps with this revision in mind, Nuttall had introduced himself to the Landreth Nursery with a letter of recommendation from the John Frasers, London nurserymen who had botanized in the Appalachians before returning to London to start their business. Nuttall had finally accepted the idea and necessity of collecting for others for money, even of lecturing to others for money. His days of setting type and accepting a meager allowance for his true work had ended, but he could never stop traveling. William Bartram helpfully directed him on a southern botanical trip in his own footsteps, and also quite possibly along the route followed by Pursh later. After a southern trip to Savannah, Nuttall undertook another down the Ohio and around Kentucky to the Carolinas. Both trips involved collecting for subscribers and verifying the discoveries of others. In 1818 Nuttall published, instead of a rehash of Pursh, his own *Genera of North American Plants*.

Nuttall's book went far beyond the works of both Pursh and Michaux (whose *Sylva* he later edited and amplified, as he had first intended to do with Pursh). Although it did justice to the studies of local flora undertaken in many eastern localities, it also moved the world of American botany farther west. Moreover, it was written in English, with full descriptions of Nuttall's discoveries. An unfair—though probably correct—criticism of the book contends that Nuttall was deficient in his knowledge of taxonomy, as, in fact, had been his instructor, Benjamin Smith Barton. A real injustice seems to have been that there was no publisher for the work; Nuttall drew on his earlier training and published it himself, setting much of the type. Rafinesque wrote the first review, in 1819, and described the work succinctly:

we find that it is not a mere description of our genera, but an enlarged survey of them. After the botanical English names of each genus follows a correct definition of it, in the style of Jussieu,

with observations on the habits and peculiarities of it. Next a catalogue of the species . . . including many new ones, of which full
descriptions are given, and lastly an account of the number and
geography of the foreign species belonging to the same genus.
Therefore the whole includes a more correct account of our genera
than has ever been published.

Caleb Cushing, appointed our first ambassador to China, wrote the
only other review, reasonably favorable but critical of literary execution and of a tendency to subdivide old genera. Magazines for review-
ing were as scarce as books to be reviewed, but the work was cordially
commented upon by the whole fraternity of botanists. John Torrey,
one of the greatest of these, who was to dedicate his forthcoming book
on regional flora to Nuttall, said the *Genera* "contributed more than
any other work to advance the accurate knowledge of plants of this
country."

Nuttall was by this time a member of the two great Philadelphia societies, the Academy of Natural Sciences and the American Philosophical Society.[5] Four friends from the latter organization donated fifty
dollars each to support his next journey, toward the Southwest, which
Nuttall felt was full of undiscovered treasure.

He set out for the Arkansas River and a voyage that lasted a year and
a half. Again he went to Pittsburgh by mail coach, but this time (his
third) he determined to ease his travel by taking a skiff and a passenger
to help. When the river became too formidable for the skiff, Nuttall
purchased a flatboat and continued, with two different passengers.
When he reached the Mississippi, the river was still full of ice and
forced a delay, but Nuttall reached the point at which he intended to go
upstream on the Arkansas River sooner than he had hoped. By skiff
again, he made his objective at Fort Smith. Later in the spring, he
joined an expedition overland. The group's purpose was to put white
settlers off lands the government had granted to the Osage Indians.
Nuttall was so beguiled by the territory that he missed the return journey, having lost his way and his horse, and was profitably delayed. He
saw *Maclura* (Osage orange), named for one of his patrons, in abundance, and a host of other Texas wildflowers now of great value to our
gardens—coreopsis, rudbeckias, pentstemons, and others. He also
saw warlike Indians, including Osages. After five weeks of trouble in

Two sketches from Thomas Nuttall's *Journal of Travels into the Arkansa Territory during the Year 1819* (Philadelphia: T. H. Palmer, 1821). Courtesy Hunt Institute for Botanical Documentation, Carnegie-Mellon University, Pittsburgh, Pennsylvania.

the wilds, he returned to Fort Smith to find that a new territory had been created by the United States government—Arkansas, now distinct from Missouri. Finally, he made his way back to New Orleans and from there to Philadelphia.

In 1821 he published *A Journal of Travels into the Arkansa Territory during the year 1819, with Occasional Observations on the Manners of the Aborigenes*. It was illustrated by five engravings drawn by the author, two of which we reproduce here. A favorable review was written by Jacob Bigelow. Nuttall read a series of papers describing "some Arkansas plants recently introduced into American gardens," the seeds for which he had already distributed to the Philadelphia nurserymen, to the Prince Nursery on Long Island, and to his subscribers in England. The system of naming his discoveries of new species by placing an asterisk between the genus and the specific name seems fair enough and very helpful, as does his referring to the little Blue-eyed Mary, which so caught his fancy with its azure lip, as *Collinsia*verna* Nutt., because he had discovered it and named it for his friend, Zaccheus Collins. The garden seeds he mentioned included: *Coreopsis*tinctoria, Helianthus*petiolaris, Astragalus*nuttallianus, Centaurea*americana, Haplopappus*ciliatus* Nutt., *Palafoxia*callosa* Nutt., *Nemophila*phaceliodes, Astragalus*micranthus, Verbena*bipinnatifida, Callirhoe*digitata*, and four species of *Oenothera*, **linifolia*, **serrulata*, **speciosa*, and **triloba*. These were only a few of the plants, which Nuttall estimated as about 300, he had brought back from the Southwest. As grateful gardeners, we may recognize some of these discoveries under the names as given by Nuttall here. Looking them up in Bailey's *Standard Cyclopedia of Horticulture* today, we are pleased to see *Centaurea americana* Nutt., the basket flower, rated "very attractive." *Verbena pinnatifida* Nutt. is still popular as a perennial "whose flowers dry bluish purple." *Coreopsis tinctoria* Nutt. is a common garden annual, "showy and good." *Nemophila phaceliodes* Nutt. is described as "sparsely hairy." Not all Nuttall's treasures still bear his name; problems in nomenclature are always being settled and resettled by experts. But his name still appears frequently in lists of American plants.

After this trip, Nuttall seemed to be on the verge of changing his life style, to begin lecturing and making friends with settled botanists like John Torrey, to start teaching. He suddenly received appointments as

Coreopsis and *Rudbeckia purpurea* (now *Echinacea purpurea*): two American wildflowers that gave a new look to border plantings in English and American gardens. From Curtis, *Botanical Magazine*.

curator of the Botanic Garden at Cambridge[6] and as instructor of natural history at Harvard. Nuttall wrote from Philadelphia to young John Lowell, son of the Botanic Garden's originating spirit, about being "unexpectedly honoured" by his appointment and said he had already forwarded two boxes of plants and "one matted bundle" for the garden,

> rare and I hope the gardener will attend to them. . . . I have obtained . . . a considerable number of seeds and have the prospect of many more. . . . I know of nothing that is wanted to establish a general collection of plants at Cambridge but industry and economy. A good botanic is yet a desideratum.

He arrived to take up his duties in 1823. Here we might expect to leave him "vegetating," as he said, because he seemed to have quit being one of our free spirits. But how little even he could guess.

Nuttall's time at Harvard was full of work and interruptions. During this period, Nuttall was deeply involved with his two-volume *Manual of the Ornithology of the United States and of Canada*. He was granted several leaves of absence—to add to the gardens; to get material for his *Ornithology;* to go to England to settle affairs for his Uncle Jonas. He initiated short field trips for his classes and became such an original lecturer that a Boston youth of impeccable upbringing was often encouraged to entertain his parents' friends after dinner by imitating Nuttall conducting a class. But this was all "vegetating."

When a young acquaintance of equally adventurous spirit, Nathaniel Jarvis Wyeth, after a successful year of exploring, colonizing Oregon, and collecting for the garden, suggested that Nuttall join him, the die was cast. Nuttall asked for yet one more leave of absence and offered to resign. His resignation was accepted on March 20, 1834.

After a great flurry of finishing papers and collecting data, visiting friends and buying supplies, Nuttall found himself again on the stage for Pittsburgh, then on a steamboat for St. Louis, and impatiently and joyfully setting out ahead of the party on foot, to be picked up later by boat. Nuttall was on his way to the Oregon Trail.

By this time, nearly fifty and still a free spirit, Nuttall was the world's leading botanist. We leave him now, ending with his own summary of his travels, as given in the preface to *North American Sylva* (his supplement to Michaux's *Sylva*), published in 1849.

> Thirty-four years ago, I left England to explore the natural history of the United States. . . . [A]fter a . . . dangerous passage, our dismasted vessel entered the Capes of Delaware in the month of April. . . . As we sailed up the Delaware, my eyes were riveted on the landscape with intense admiration. All was new; and life, like that season, was then full of hope and enthusiasm. The forests, apparently unbroken in their primeval solitude and repose, spread themselves on either hand as we passed placidly along. The extending vista of dark Pines gave an air of deep sadness to the wilderness:—
>
> . . . For thousands of miles my chief converse has been in the wilderness with the spontaneous productions of nature; and the study of these objects and their contemplation has been to me a source of constant delight.

This fervid curiosity led me to the banks of the Ohio, through the dark forests and brakes of the Mississippi, to the distant lakes of the northern frontier; through the wilds of Florida; far up the Red River and the Missouri, and through the territory of Arkansa; at last over the "Vast savannas, where the wandering eye, / Un-fix'd, is in a verdant ocean lost"; And now across the arid plains of the Far West, beyond the steppes of the Rocky Mountains down the Oregon to the extended shores of the Pacific

and on, to the Sandwich Islands, which we will omit here, until his return:

Leaving this favored region of perpetual mildness, I now arrived on the shores of California, at Monterey. The early spring (March) had already spread out its varied carpet of flowers; all of them had to me the charm of novelty, and many were adorned with the most brilliant and varied hues. The forest trees were new to my view. A magpie, almost like that of Europe (but with a yellow bill) chattered from the branches of an Oak with leaves like those of the Holly (*Quercus agrifolia*). A thorny Gooseberry, forming a small tree, appeared clad with pendulous flowers as brilliant as those of a Fuchsia. A new Plane Tree spread its wide arms over the magnitude of a small tree, loaded with sky-blue withered flowers, [which] lay on the rude wood-pile, consigned to the menial office of affording fuel.

After a perilous passage around Cape Horn, the dreary extremity of South America, amid mountains of ice which opposed our progress in unusual array, we arrived again at the shores of the Atlantic. Once more I hailed those delightful scenes of nature with which I had been so long associated. I rambled again through the shade of the Atlantic forests, or culled some rare productions of Flora in their native wilds.

THE tales we have recounted concern only a few of the really free spirits, who followed their own ways to their own peril and pleasure. A host of others wandered and searched, collected, and sometimes lost—even themselves. Because we are working on a small scale, in gardens, we must leave out the great explorers who collected the trees that grow today in our vast parks and arboretums. The stories of all these discoverers make splendid reading, and their deaths seem unbearable news even today. We shall use the flowers they found, checked, and confirmed, as we fill our borders.

It is important to understand here that the mid-nineteenth-century American garden, whether attached to a mansion or a cottage, was primarily by force of circumstances a botanist's garden. Walter Elder, a somewhat ignorant but nonetheless dictatorial writer who concerned himself with the "*intelligent* cottagers' gardens" in midcentury, makes a very neat distinction between types of plant lovers and their effects on plants.

The Botanists and the Florists

The botanists and the florists are distinct persons, and their theories are quite opposite to one another. The botanist delights in nature—the florist in art. The botanist is at home in all parts of the world where plants are in bloom—the florist's world is the flower garden. The botanist is amused with the stamens and pistils—the florist with the flower leaves [petals]. The botanist takes a plant to study its structure and nature—the florist takes a plant to triple the size and number of its flower leaves. The botanist considers a plant with a double flower a monster—the florist considers it a beauty. A double flower is useless to the botanist—it is a prize to the florist. The botanist loves to see plants in their natural characters—the florist loves to see their characters altered by hybridization and extra culture. Species are the hobby of the botanist— variation the hobby of the florist.

The most ornamental of the blooms that had so charmed the free spirits who sought them across the prairies went straight into the gardens without awaiting "improvements." The little "azure-tipped" treasures of the botanists received royal welcomes and began long lives as garden favorites. After the identifying and cataloguing of their lesser selves and earliest appearances by the fixed stars like Gray and Torrey and many, many others, the future florists' worlds of unproved zinnias and dahlias, sweet peas and hyacinths, and even roses and lilies, would invade the nineteenth-century gardens and fill them literally to overflowing where the sturdy little originals with azure lips were staunchly holding their own.

NOTES TO CHAPTER 4

1 Rafinesque reported his ambition to Jefferson and begged to be included on any expedition Jefferson was then contemplating. Later, he importuned Jefferson for a position teaching any, or even all, sciences at the University of Virginia.

2 In Rafinesque's *Natural History of the Fishes Inhabiting the River Ohio and Its Tributary Streams.* In this book, Rafinesque followed a procedure he had developed in Sicily, of presenting his fish, like birds, in glowing colors.

3 We may be reminded of the London Quaker merchant, Peter Collinson, who urged our "natural botanist," John Bartram, to journey into fresh country where, Collinson felt, hostile Indians could be won over by the friendly overtures of solitary wanderers. Even when Bartram protested that one Indian he had met had torn Bartram's hat from his head and chewed the brim, as if to show what he would like to do to Bartram himself, Collinson chided him for lack of faith.

4 Better for us to read about it in *Astoria,* by Washington Irving, and in the brilliant accounts of Nuttall's explorations in America from 1818 to 1841, described by an expert plant lover, Jeanette E. Graustein, in her book, *Thomas Nuttall, Naturalist.*

5 The president of the American Philosophical Society at this time was William Maclure, author of the first book on the *Geology of the United States* and a philanthropist, whose interests would include Nuttall's Arkansas trip and his own involvement with the experimental settlement in New Harmony, Indiana.

6 This establishment was the inspiration of John Lowell, president of the Massachusetts Society for Promoting Agriculture. His sons, John and Francis, carried on his plans after his death. William Dandrige Peck (remembered today for his discovery in the White Mountains of *Geum peckii* Pursh) was the first to occupy the position of professor of natural history and worked hard to set up the garden on land donated in part by Andrew Craigie (whose house later became Longfellow's—the site of a garden described later in this book). A house similar to Craigie's, although smaller and more elegant, was built in the garden for the curator-cum-professor. This is the garden to which Asa Gray, like his friend John Torrey, succeeded before he published his *Manual of the Botany of the Northern United States* in 1848. It existed as a garden of long, narrow, raised beds well into the twentieth century, giving its name to sections of Cambridge called Gray Gardens East and Gray Gardens West, now totally given over to housing, leaving only a herbarium to carry on the tradition of the first academic botanic garden.

Seedsmen and Their Nurseries

ROM the turn of the century, beginning with the impressive contribution of nurseryman Bernard M'Mahon, there was a crowding of knowledgeable men eager to share their experiences and sell their plants. They seem to have formed a freemasonry of experts, starting first on the Atlantic seaboard and advancing west, like the gardens and orchards they fostered. Some came from abroad, others descended from early settlers who had come by their local knowledge the hard way. Some founded families who carried on their work on the original sites. They were competent, sincere, friendly, and able to back up their advice with living plants and trees. And to a man they wrote.

This was just as well, as the possible market was enormous. The prospective customers were literate, and the distances were prodigious. New discoveries, useful tips and hints, procedures in preparation—all could fade and disappear before they could be discussed and shared. One successful experiment could stimulate many more, provided it could be discussed. Seeds saved from some successful experiment would be at a premium as soon as the news got out. In a new country with unlimited natural resources, even the time-honored plants from the Old World could take on a new life. Even the old fruits—especially apples—could be born again to produce miracles. And the Old World waited eagerly. Organization and communication were all that were needed.

In the century before, we saw the Bartrams, father and son, near Philadelphia, carrying on explorations for unusual native flowers and trees (encouraged by "curious" gardeners abroad—and often in their pay) and keeping a nursery with a catalogue list from which people like George Washington ordered. There was also the Prince Nursery on Long Island, so successful and well known that it was carefully spared by both sides during the Revolution. The Prince Nursery got out enormous catalogue sheets annually, developed into a family business, and later took to the printed word (on grapes and roses).

None of these earlier nurserymen had set forth a thorough body of information until *The American Gardener's Calendar* by Bernard M'Mahon appeared in 1806. Published in Philadelphia, this book was the first to offer professional advice to those gardening on American soils and with the vagaries of the American climate. It had come a far distance from previous centuries of reliance on the wisdom of the "antients," Virgil and Columella, who had guided agriculturists through the previous centuries.

M'Mahon's impressive first sentence packs in his intentions, while demonstrating his view of the specifically, newly democratic, American setting.

> The general utility of Horticulture, or the art of improving every kind of soil, or producing a plentiful supply of wholesome vegetables and fruits, so necessary to health in all countries, especially in warm climates; of cultivating the various plants designed by Infinite Goodness to minister to the comforts of animal life, by correcting the divers maladies to which it is subject by nature, and still more so, in the human race by intemperance; as well as materials for ornamenting the whole face of the country; it is too obvious to render any arguments necessary in favour of an attempt to facilitate the general acquisition of that useful branch of knowledge; but more especially in a country which has not yet made that rapid progress in Gardening, ornamental planting, and fanciful rural designs which might naturally be expected from an intelligent, happy and independent people, possessed so universally

of landed property, unoppressed by taxation or tithes, and blest with consequent comfort and affluence.

M'Mahon continues disarmingly to attribute neglect "in these respects" to a lack of pertinent information, because all that was then available came from foreign countries of differing climates, soils, and situations. It is for the welfare of his fellow citizens and the general improvement of the country that M'Mahon is undertaking his work.

And a great work it is, we realize now, perhaps even as a voice in the wilderness. M'Mahon must have been a winning personality as well as a sound gardener, with experience "of nearly thirty years in PRACTICAL GARDENING." His instructions are divided into months and are related to the American climate. He includes a general catalogue of over 3,700 species and "varieties of the most valuable and curious plants hitherto discovered." With this book in hand, if "the reader has a taste for admiring and enjoying the magnificence, beauties and bounties of nature in its vegetable production," he may, after a few years "of daily study and amusement," become a "Complete Master of the Art, and if he pleases, his own Gardener." Considering that there are chapters on the care and feeding of silkworms, the cultivation of vineyards, the proper crops for feeding stock, and the making of live hedges, M'Mahon's hopes for the absorption of his store of knowledge in less than a lifetime may have been overly sanguine, but his sincerity and intention to be of service to his new country are clear.

Echoes of previous centuries are implicit in his general summary— the idea that "Infinite Goodness" had designed the plant world was understood in the first settling of the country. Hints of luxury and commerce as incentives were woven into the lives of those who succeeded the Puritans. But the idea of "ornamental planting, and fanciful rural designs," not to mention "ornamenting the whole face of the country," is totally nineteenth-century, first put forth in print by M'Mahon.

Bernard M'Mahon was one of very few innovationists on the American scene to be contemporaneously recognized and rewarded. His book became the authority on how to succeed in gardening and enjoy it, too. For years, edition after edition was respectfully consulted. That later carpers (and they do spring up, like weeds, in all the choicest locations) questioned M'Mahon's ability to supply, or even to have seen,

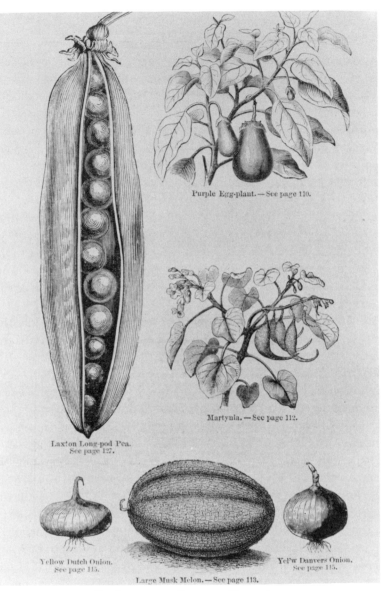

Purple Egg-plant. — See page 110.

Martynia. — See page 112.

Laxton Long-pod Pea.
See page 127.

Yellow Dutch Onion.
See page 115.

Large Musk Melon. — See page 113.

Yel'w Danvers Onion.
See page 115.

Catalogue page for vegetable seeds.

many of the plants he so expansively recommended is beside the point. He did stimulate an interest in new gardens on new soil with new plants and placed their pursuit within the reach of anyone eager to begin "ornamenting the whole face of the country" by concentrating first on his own back yard.[1]

A dynamic English farming authority, William Cobbett, must have done well for himself with the many editions of his little book, *The American Gardener.* The copy we have is London, 1821. Unlike M'Mahon's great work, however, it is a treatise on gardening in England with frequent admonitions to remember we are on Long Island. The book is dedicated to Cobbett's American hostess, who was especially fond of flowers. In her honor, a rather skimpy sixth of the book is devoted to "Flowers and Ornamental Gardening" in general, and the plants listed and described include also trees and shrubs. The book's popularity may have been due to the frequent reminders that all the "finest trees and shrubs in England" came from America.

M'Mahon, with his friendly approach to business, was soon followed by a throng of able businessmen, intelligent, literate, and expert gardeners. Some of them had arrived on American shores already trained in horticultural matters, either by apprenticeship in botanic gardens in the Old World or by having managed large estates. Some of the most successful of our recognized authorities started as head gardeners to ambitious landowners close to Philadelphia and up the Hudson. In George Washington's day, German gardeners had been much sought after. With the nineteenth century and the great advancement of garden practices in England, Scottish horticulturalists were most popular in other parts of the British Isles and in America, and they were at a premium. Perhaps the harshness of their native climate contributed to their ability to handle American extremes of weather and soil conditions, which would have dismayed the English and which still startle English visitors.

In the correspondence columns of Loudon's enormously influential *Gardener's Magazine,* advice was sought from those already here and given to those impatient to leave the tedium of gardening-under-orders for the joys of independent management. But encouragement from those already here was wisely lacking. Gardening was hard work in the New World, too. After the heated correspondence in the *Gar-*

dener's Magazine, aspiring gardeners seeking to migrate could not complain they did not know what they were getting into.

In 1831, William Wynne, the foreman of Bartram's Botanic Garden, wrote to J. C. Loudon describing the "state of horticulture" in and around Philadelphia and commenting on the advisability of British gardeners' emigration to the United States. He first sketched the general scene as he saw it, writing of the preeminence of the Bartram garden—the oldest by a hundred years and the richest in American plants. He added:

> This garden is the regular resort of the learned and scientific gentlemen of Philadelphia. . . . A committee of the Horticultural Society closes an account of this nursery as follows. . . . "Mr. Carr, who deserves so much credit for the classification of his nursery, is no less entitled to praise for the admirable order in which his toolhouse is kept; a place that, in most gardens, instead of possessing regularity, is made a mere lumber room. The best order is likewise preserved in the seed room, in putting up our native seed. That apartment, moreover, contains a library of 400 volumes, in which are all the late works on botany and horticulture."

Wynne continued by pointing out that the next-best nursery to Bartram's, in extent and variety, is one kept by Messrs Landreth. Mr. Wynne gives them credit for "a good stock of greenhouse plants, orange and lemon trees loaded with fruit and a remarkably fine Champney's rose."[2] Mr. Wynne also reports on the small nursery of a Mr. Hibbert, whose roses are sold in pots in the city market. Mr. Hibbert had recently bought a piece of land formerly occupied as a nursery by Mr. M'Mahon and had taken into partnership Mr. Buist, a gardener in the neighborhood. He also notes that

> There is another class of gardens, very distinct from any I have seen before: those of plant-growers, who, to a small nursery and green and hot houses, add the appendage of a tavern. The two principal ones of this description are kept by Mr. Arran and M. d'Arras: the first has a very good museum in his garden; and the latter possesses a beautiful collection of orange and lemon trees, very large but trimmed after the French fashion. These

places are the resort of many of the citizens; Philadelphia having no parks, or national gardens, for the purpose of recreation.

Despite his general positive tone about the gardens, Mr. Wynne is not given to cheerful prognostications when he turns to his true subject, the prospects for the gardener who emigrates. He surmises that the newcomer will find "difficulties in his way at first, from the spirit of rivalry which everything British creates among the vulgar here." He believes that "for constant and well-finished work, and gardening, too . . . an American is as much inferior to an Englishman, as a Choctaw is to the former." In fact, a man "who can procure a good situation in Britain if he is fond of his profession should not come here." For there are, says Mr. Wynne, "no American gardeners except amateurs." And he, too, cannot help remarking that the law of primogeniture is the best friend of popular gardening. The laws in the United States work to the loss of splendid examples of sustained horticultural achievement. But he closes his letter by assuring Loudon that, if he visits "this land of freedom and plenty," his readers here will be highly delighted.[3]

This article by the acerbic Mr. Wynne is immediately followed by a most generous one by Alexander Gordon, who toured the American scene in 1827 and 1828. He praises the establishment of seedsmen like Thorburn and Floy in New York (whom we know were resorted to by Dr. Kirtland of Ohio) and regrets not having sufficiently visited several others whom he, however, kindly lists. He does justice to the Prince Nursery and to Bartram's, writes a charming account of visiting Hyde Park with Dr. Hosack, which we quote elsewhere, and takes a rewarding trip to the South, where he waxes enthusiastic over the charms and horticultural accomplishments of Charleston and Savannah, although he met "gentlemen in the south who never saw a cauliflower."

Thomas Fessenden, editor of *The New England Farmer,* portrays the more favorable side of emigration, suggesting that gardeners in the New World enjoy the "circumstances of the American farmer . . . generally the owner, as well as the occupier, of the soil which he cultivates . . . the product of his labors will command more of the necessaries, comforts, and innocent luxuries of life, than similar efforts would procure in any other part of the globe."

ANY survey of what plants were most popular in nineteenth-century American gardens depends upon the catalogues and instructional publications of these noted resources, many of whom did respond to the sentiments of Fessenden rather than to those of Wynne. Because we are dealing chiefly with gardens for flowers and ornamental trees and shrubs, we have selected as our sources a group of American nurserymen who due to circumstances are mainly from the East Coast and the first half of the century. As the population moved rapidly westward, first on foot and by ox carts, later aided by boats on lakes and rivers and much aided by the Erie Canal, and finally by train across the continent, nurserymen inevitably followed the homesteaders. But in the western regions they contributed most heavily to fruit growing and field crops.[4] Most of them left the cultivation of ornamental materials to those already entrenched in the East. But the pattern we see held true—nurserymen and seedsmen founded horticultural and agricultural and pomological societies wherever they put down. Our debt to them is still growing.

The benign visages of the group we have chosen would grace a temple like the ponderous garden ornament at Stowe honoring "British worthies." Their portraits hang high and dimly today in halls of the horticultural societies they helped to start and in the horticultural libraries to which they left their books. We present them in chronological order, depending upon the dates of appearance of those books upon which we have based our appendix of plants.

The first resource is the only woman listed under "horticulturists" in the 1950 edition of Liberty Hyde Bailey's *Cyclopedia*. She is Mrs. Annie L. Jack, who arrived in Canada from England by way of New York. Her handbook, *The Canadian Garden,* was a primary source and will be dealt with more fully when we consider Canada.

Joseph Breck of Boston is the only early American horticulturist whose business continues under the family name. Born in 1794, he published his first book, *The Young Florist,* in 1833 and his most important, *The Flower Garden; or, Breck's Book of Flowers,* in 1851. This was reissued in 1866, when he was seventy, as *The New Book of Flowers.* All three works went through numerous editions. Breck also edited *The New England Farmer* until the advent of *The Horticulturist,* under the editorship of Andrew Jackson Downing, when Breck made over his

Horticultural Hall, where seedsmen in Boston ran rival businesses and met for friendly discussions.

list of subscribers to the new publication. He joined with Thomas Fessenden to edit *The Horticultural Register* in 1836–38. An original member of the Massachusetts Horticultural Society, he was its president from 1859 to 1862. He wrote easily and generously with many personal comments, a blessing in an age where formal diction was customary.

Thomas Bridgeman arrived from England in 1824 to conduct a seedsman's business in New York under the name of his son, Alfred Bridgeman. In 1829 he began to write *The Young Gardener's Assistant,* which was many times printed and enlarged and was copyrighted in 1847. The work had been designed in three parts, concerning vegetables, fruits, and ornamental gardening. Eventually two parts were issued as *The Kitchen Gardener's Instructor* and *The Florist's Guide*. The editions we have before us are those of the three-volumes-in-one of 1845 and 1853, the second slightly more elegant than the first and having a gardener's calendar added to the original work, which, he says in the preface, he had undertaken to

> enable our respectable seedsmen, while furnishing a catalogue of seed for the use of the Kitchen and Flower Garden, to afford instructions, at a trifling expense, to such of their customers as had not a regular gardener, and thereby save themselves the blame of those who may not have given the seed a fair trial, for want of knowledge how to dispose of it in the ground.

The book was enthusiastically reviewed in the *Magazine of Horticulture, Botany, etc.*, published by Hovey and Company of Boston, who commented that "its cheapness should place it in the hands of all new beginners."

Bridgeman's work is characterized by a pervading kindliness. The plant lists are exhaustive. There are separate monthly calendars for vegetables and for flowers. The descriptions bloom with poetry. Of all the books of then or now, this one would be most pleasant to follow. For our appendix of plants we take all our herbs from this friendly work.

Robert Buist was born near Edinburgh in 1804 and trained at the Edinburgh Botanic Garden. We shall lean on Buist for our list of roses in the Appendix. He arrived in New York in 1828 to be employed by Henry Pratt, but within a very few years he entered into partnership with the firm of Hibbert, first notable florist in Philadelphia. An expert gardener and propagator, he enjoyed great success with roses and introduced an improved verbena, which seems so to have encouraged gardening at that level that he was able to introduce a whole class of bedding plants. He marketed for the first time *Poinsettia pulcherrima* and produced a double form.

Buist wrote three books, *The American Flower Garden Directory* in 1832, *The Rose Manual* in 1844, and *The Family Kitchen Gardener,* copyrighted in 1847. Our copies of the *Directory* are the second edition of 1839 and that of 1858, in the preface of which he hints briefly at the not yet proven idea of the benefit of electricity to vegetation and advises caution in the use of guano (although he notes that strong-rooted plants can take watering with a solution of one pound to five gallons of water). He writes from the Rosedale Nurseries, recommending his "table of soils," and speaks enigmatically about the estates of the wealthy, which he considers to have room for improvement. For, although the advice of European publications on gardening can prove to be a dead letter when practiced in the United States, it is not so with "their architectural and horticultural designs." Buist makes a polite bow to Loudon when he claims that American estates "want more of the picturesque and (to use the words of the veteran pioneer of horticulture) gardenesque effect, to relieve their premises of the monotonous erections and improvements which seem to govern all." In some of their undertakings, however, Buist considers the Americans second to none, as witness the success of the *Victoria regia* water lily at Spring Brook, residence of Caleb Cope, Esq.

Thomas Meehan, originally an employee of Robert Buist before becoming superintendent of Bartram's gardens, was cultivator of the *Victoria regia* lily for Caleb Cope. Editor of the *Gardener's Monthly* for thirty years, nurseryman, explorer, discoverer of the relationship between the movements of plants and glaciers, author of *Native Flowers and Ferns of the United States,* and one of the men responsible for starting the Pennsylvania Department of Forestry, he cuts across all our fields like a shaft of light.

Peter Henderson published his *Gardening for Profit* in 1866, the year he started his seed business. It ran into three editions, which revolutionized American horticulture with advice on gardening on a large scale. He followed it with *Henderson's Practical Floriculture* in 1869, a book that shook the horticultural world again with advice on commercial growing of flowers. But he did not neglect the amateur gardener, publishing *Gardening for Pleasure* and *Handbook of Plants*. All his books are direct, succinct, and, to an amateur, faintly discouraging. Perhaps we feel dragged through so much in so few pages—from lawns and insects to hanging baskets and cold graperies to "parlor gardening" (or the cultivation of plants in rooms) and warnings about the "law" of color in flowers (which, if we heed it, will save us from buying what is advertised as a "blue" rose; as soon, Henderson says, expect our poultry to assume the azure hues of our spring bluebird). He

Three plants, introduced
from China, which
became popular:

Gingko biloba Linn. (Engelbert Kaempfer's
Amoenitatum Exoticarum, 1712).

also knows about roses. For us, however, he will demonstrate the plantings of flower beds and be our mainstay in the Appendix, with his listing of familiar names, which he had made with "great care." He also endears himself by acknowledging his indebtedness to the seventeenth-century's John Gerard and John Parkinson, as well as to Loudon, Paxton, and Gray.

Besides these practical offers of sound advice to average home-owners, there were other guides to the world of American horticulture. Thomas G. Fessenden, remembered today chiefly for his satirical verse (unfortunately), was extremely influential with his books, published early in the century: *The Complete Farmer and Rural Economist, The New American Gardener,* and *The American Kitchen Gardener.* Criticized as being too English in form and for referring very little to growing plants, he is, nevertheless, our source for a copy of Parmentier's article on gardening.[5] In his way, Fessenden sparked the brush fire of books on gardening with his "perspicuous and practical sketches of

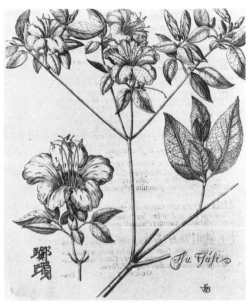

Camellia japonica Linn. (Kaempfer, *Amoenitatum Exoticarum*, 1712).

Azalea (now *Rhododendron*) *kaempferii* var. *japonicum*. (Kaempfer, *Amoenitatum Exoticarum*, 1712).

some important improvements in modern husbandry." He also, as we have noted, encouraged experienced gardeners to try their hands in the New World.

The Hovey brothers, Charles Mason and Phineas Brown Hovey, represent a new powerhouse of ability turned to public benefit. They established a nursery in Cambridge, Massachusetts, in 1832, having long worked toward such an enterprise. In 1834 Charles Hovey introduced a strawberry that revolutionized American fruit culture, its success stimulating experiments in all other sorts of fruits. A magazine was started by the brothers, with a name and frontispiece so blatantly imitating Loudon's *Gardener's Magazine* that the format could be attributed kindly only to outright flattery. There may have been less charitable comments, for, within a few copies, *The American Gardener's Magazine and Register* became *The Magazine of Horticulture, Botany, and All Useful Discoveries and Improvements in Rural Affairs*, with Charles Hovey as editor, and so it remained until 1868.

Hovey was greatly praised by the nurseryman Thomas Meehan, when Meehan announced in 1886 in his *Gardener's Monthly* that "horticulture on this continent is probably more indebted to Charles Mason Hovey than to any living man." To us today, Hovey's major importance lies in his great experiments in plant hybridization. After his initial success with the strawberry, he undertook to hybridize other fruits—the Hovey pear and the Hovey cherry—as well as ornamental trees and shrubs, such as camellias and azaleas. He may be the most important nineteenth-century seedsman of all. He died in 1887 and was buried in Mount Auburn Cemetery, at the corner of Mound and Spruce.

WALTER ELDER, in his contemporary comments in his little book, *The Cottage Garden of America,* published in Philadelphia in 1848 with a second and improved edition in 1854, has done us great service by listing the "American Authors on Horticulture" of his time. He allows Robert Buist to head the list, with Thomas Bridgeman to follow, and then A. J. Downing. His order would appear to be alphabetical, except that he tosses in several worthies at the very end: Fessenden, Nuttal (*sic*), and Barton. As Barton's *North American Flora* "is said to be a most splendid work," we may assume the author has not read it but felt bound to mention it. He says "Nuttal's *Genera of American Plants* is a useful book," and Fessenden's writings are "valuable," in common with the works on fruits by Sayers and Thacher. In fact, when we study his list—beginning with Buist, Bridgeman, and Downing— those three seem to be the authors Elder has read. He praises Floy, whom we are glad to see mentioned, as he figures largely in the orders of Dr. Kirtland of Ohio. "Messers Hovey, seed and nurserymen of Boston," he says are "very scientific." David Landreth is "pleasing." Mr. Kenrick of Boston has written a superb book on *The American Orchardist.* Mr. McMahon's book is "good." Mr. Prince of Long Island wrote a "valuable" book on the vine. Mr. Downing is listed again as editor of a monthly periodical, *The Horticulturist.* And there we have it, according to Mr. Elder.

His list of ready reference books seems mainly to be credited to East Coast "seed and nurserymen," which helps us to feel we have not withheld credit from contributors farther west, where the gardens and gardeners—if not the authors—were all heading.

Elder's grouping of these men has another bracket for his purposes, however, and not a laudatory one, for Elder feels they all address "the inhabitants of the *mansion*." "Some of them may say 'how do you do' to the cottager at a distance, but they then pass on seemingly afraid to be thought associating with him. So we have taken untrodden ground in the field and address ourselves entirely to the intelligent *cottagers* of America." He can conceive of the relationship of landlords and tenants in connection with his cottage gardens, and he lays down duties for each. First of all, a landlord will "benefit" from a happy and healthy tenant, in a house on slightly rising ground, neatly fenced in, with a grass plat for a "bleaching ground" and sturdy poles deeply set to carry a clothesline. Gardens are important, as they are "reforming and moralising to the young" and "exalt the national character." Second, tenants' duties are to take care of the property, to make no complaints, and to stay in one place as long as possible, because they will then be "happy in themselves," get the respect of their neighbors, and find their landlords generally willing. Elder does not feel, with others of his time (particularly F. J. Scott whom we meet later), that it is up to either landlord or tenant to sacrifice the front half of a property for an unfenced area to be admired by passing strangers. Elder feels that a well-kept cottage garden in itself is both "patriotic" and "christian."

Because Elder feels so strongly for the "intelligent cottager," we have used his advice as a final reference to roses in our Appendix, following all those other "seed and nurserymen" of whom he approved, despite their tendency to address "the inhabitants of the *mansion*."

NOTES TO CHAPTER 5

1   And here a word about *yard*. The term came from a very old English source and in America came to mean the area in front, on the side, or in back of one's house—or, as Americans liked to say when they settled in for life, "home." The garden was what one planted in the yard—the vegetable garden in back, a flower garden in front or on the side. The dooryard was obvious. Visitors found, and still find, the nomenclature quaint, but so it is.

2   This last is of interest, as it is the first successful crossing of the everblooming China rose with the once-blooming European rose. It is good to see it called "Champney" for its propagator in Charleston, South Carolina, before it disappeared into the Noisette roses forever.

3 Little can Mr. Wynne suspect the recreational joys awaiting Philadelphians in landscaped cemeteries, even one made from the great estate of the Woodlands, so lately worked in by such worthies as the senior Landreth and Mr. Pursh.

4 As the new regions were opened up and virgin soils were literally deflowered, American fruits—runaways from those brought by the early settlers—created a whole new world of their own. Concentrating in the new species, and later the hundreds of varieties, could absorb the attention and fill the purses of all the nurserymen willing to chance their fortunes in what had been wilderness. We see them advancing westward across the country simply by reading lists of those fruits most prominent in the nineteenth century.

5 Printed in *The New England Farmer,* which he edited, one of only two subscription papers on agriculture at the time. The other was *The American Farmer,* published in Baltimore.

Enter the Ladies

HE nineteenth century in France and England—and inevitably in the developing New World of North America—was imbued with the determination to disseminate knowledge. In American life in the early nineteenth century, with the beginning of popular education, common schools, and public libraries, the very act of learning was respected. To be said to have educated oneself was a cause for envy rather than commiseration. To hand on what one had learned was the backbone of every one-room country school.

Certain subjects seemed of universal and immediate usefulness— botany, for instance. Beyond studies and accounts of rival scientists, adventurers, and explorers, there was an urge to introduce the subject—suitably tempered and watered down to accommodate their weaknesses—to children and to ladies.

Children were considered able to take botany head-on, in textbooks written especially for common schools by such worthies as the leading botanist of the day, Asa Gray. By the middle of the century there was a growing spate of editions of the small book Gray wrote for children. The serious intent of the work can be gauged by a glance at the cover, which remained the same throughout the nearly countless printings. Framed in an elaborate rustic arch festooned with blossoming vines—

Two facsimile pages from Asa Gray's *How Plants Grow.*

honeysuckles, roses, morning glories—is a scene of industry, with garden tools abounding, conducted by three boys in long trousers, rolled-up shirtsleeves, and hats, and a girl in a hat and a long skirt. They are grouped before a cupola-topped, verandaed residence backed by tall Norway spruces. The title of the work is entwined on banners along the floral border: "Botany for Young People . . . How Plants Grow . . . Popular Flora . . . Of Common Plants," with the author's name relegated to lie in perspective on the stepping-stone—"Prof. Asa Gray, M.D." With all this reassurance—the ideal American home surrounded by familiar points like popular spruces and the three commonest "draperies" over the rustic arch—what could be more heartening than to know a doctor is in charge of the entire operation?

The first page, with its surround of the three most popular lilies—the Canadian, the martagon, and the little lilies of the valley—continues the tone of helpfulness. A quotation known to everyone old enough to go to church—which meant everyone who could sit still—introduces the chapter with an admonition to consider the lilies. The author explains that "Our Lord's direct object in this lesson of the Lilies was to convince the people of God's care for them." One of the ways is clothing the earth with plants and flowers, which we are to consider with attention. So we begin. And with the author's assurance that, after we have "thoroughly mastered this little book," we will be prepared to study the author's other books—the *Lessons in Botany and Vegetable Physiology* and the *Manual of the Botany of the Northern United States*—if we are very young, we are on our way.

Ladies, on the other hand, apparently were regarded as creatures who must be coaxed into acquaintance with the natural world. For them, sentiment and poetry were enlisted to make botany palatable.

There were honorable and ancient antecedents to assisting learning by conveying information in verse. The *Georgics* of Virgil, in which the basic truths of agriculture had been versified nearly 2,000 years before, had still been regarded as a reliable reference for ardent farmers like Washington and Jefferson. Early settlers in the seventeenth century felt that information in verse was easier to take than straight prose exposition. Even Governor Bradford of Plymouth had written the history of his colony's development in verse—bad verse, but verse. However, special attention to the susceptibilities of ladies was new.[1]

An outstanding example of teaching cloaked in poetry was impressively presented in England at the end of the eighteenth century in *The Botanic Garden* by Erasmus Darwin. The first American edition in 1798, taken from the British fourth edition, is heavily fortified with notes. Grandfather of the man who was later to shake the century, the author aims "to Enlist Imagination under the banner of Science . . . [and] to induce the ingenious to cultivate the knowledge of botany by introducing them to the vestibule of that delightful science and recommending to their attention the immortal works of the celebrated Swedish Naturalist, Linnaeus."

We can sense that our author, on the verge of his grandson's world of scientific documentation, is unable to forgo drawing upon the wealth of his own world of classical myths. For instance, he insists that "the Rosicrucian doctrine of Gnomes, Sylphs, Nymphs and Salamanders" can afford "proper machinery" for a botanic poem and that "important operations of Nature" can be "allegorized in heathen mythology." Having given himself license to use all the poetic props available, he begins the first of four cantos with a dramatic scene of "the Genius of the place," inviting the Goddess of Botany to descend and be received by Spring and the Elements. With the follow-up address to the Elements, we are in for a deluge of rare and arresting references: to the alchemical prowess of Bacon, whom the elements taught to explore metallic veins; to an unfortunate Mr. Richardson, who was experimenting with lightning and was struck dead; and, finally, to those elements who

> led your Franklin to your glazed retreats
> Bade his bold arm invade the lowring sky
> And seize the tiptoe lightnings as they fly.

By the time we reach the second canto, we are prepared for Gnomes, though perhaps not for all they bring to the mind of the author: Adonis, St. Paul, Venus, precious stones and metals, slavery, Hannibal. The third canto brings us tides, mermaids, the Nile, caravan drinking, the untimely death of Mrs. French (a neighbor), a monument for Mr. Brindley (another), and a pretty allusion to "Derwent's willowy dells," where the author lived. It ends with an exhortation to mothers to nurse their babies. The fourth canto addresses Sylphs and deals with trade winds, the crocodile and its egg, silkworms, Aaron's rod, hummingbirds, Kew Gardens, and the royal family. After an offering to Hygeia, the Goddess of Botany departs—followed by voluminous notes.

The second poem, which we know will be in another vein, begins with a bit of prose breaking down the Linnaean system for us. Quite simply. Twenty-four classes are divided into 120 orders, in which are 2,000 families or genera, and 20,000 species with "innumerable Varieties." And there is a simple key to the whole, an "ingenious system" by which classes are distinguished by the "number, situation, adhesion or reciprocal proportion of the males in each flower." Holding firmly to that key, we turn it again to find that "Orders . . . are distinguished by the number or other circumstances of the females." Another turn, and we are told the families or genera are known by the analogy of all the parts of the flower or fructification. Species are distinguished by the foliage of the plant, and varieties by any accidental circumstance of color, taste, and odor.

Firmly locked into this armor, we can proceed to the flowery dreamland where

> Beaux and Beauties crowd the gaudy grove,
> And woo and win their vegetable loves.

A sentimental wonderland, even melodramatic, where the violet is "lovesick," the cowslip "jealous," the snowdrop "cold," and the virgin lily given to "secret sighs."

"Vegetable loves" celebrates the confusion of science with sentiment, and vice versa, that distinguished the early nineteenth century until such poetic playfulness was thrown to the winds by the revelation of the origins of all species, including our own.

With the beginning of the century, however, the first gentle breezes of spreading affluence, domestic tranquillity, parlor needlework, inexpensive labor, leisure for self-improvement, and the charms of polite learning were vouchsafed to many. There seemed to be no art that could not be personally mastered—and even overdone. Architectural details blossomed in wood and stone and brick, often in all three combined. Gardens flourished in ribbon bedding, piles of planted rocks, and dramatically contorted trees. Literary effusions were entwined in horticulture.

Small wonder, then, that parlor games and indoor social pursuits were able to embrace the new interest in flowers, per se.

IN parlors where females of the family would sit for hours, making wool into most unlikely flowers while their other halves read aloud, explaining difficult passages, the concept of a "language of flowers" was welcomed—even when, as in the case of Darwin's poem, it involved first a firm grounding in the principles of botany.

While ladies embroidered and painted flowers and tinted engravings of bouquets in magazines and books, while children pressed blossoms, and while cooks candied them, what could evolve more naturally than the idea that messages could be sent through flowers, sentiments expressed, answers given? All that was needed was a convenient garden and the sense to realize that message senders had better be using identical reference books. For, in a time when modest fortunes could be made by writing prettily illustrated popular instructions on how to do practically anything with nimble fingers and a needle or jigsaw, runaway imaginations could lead even to falsification of established meanings. In the days when to be shocked was a sign of superiority, one could not be too careful to check one's authorities. The language of flowers offered obvious pitfalls to all but the most timid toes.

That "ladies" were usually the acknowledged authors of the little books that began to rest upon crocheted covers or velvet cloths on parlor tables was, unhappily, no insurance of their reliability. As greedy ambitions soared, so did the number of flowers and their messages. From simple expressions of love and acquiescence, some books knew no bounds. Duels could have resulted between protagonists using different sources.

Flora's Lexicon, for instance, by the poetic Catherine Waterman, published in Philadelphia in 1840 as an *Interpretation of the Language and Sentiment of Flowers* and including an Outline of Botany," could have held the field alone. Rather like a home medicine book, it has two indexes of over 200 items each—one of the familiar names of the flowers and trees dealt with, one of the sentiments capable of being expressed. And, of course, a thorough exposition of the Linnaean system, including explanation of the possibly confusing Greek and Latin terms. As a further guarantee of worth, every entry has a poem or quotation to illustrate what the compiler means by her attribution.

Better to exchange duplicate copies of such an earnest exposition than to entertain also the gaudier examples that began to pour forth. For instance, a neat, small "Handbook" of the "Sentiment and Poetry of Flowers," published in Boston in 1845, has poems to the flowers by a wide range of talents but no short cut to botany.

Unfortunately, the sentiments expressed by particular flowers in these two American offerings are embarrassingly dissimilar. The yellow rose in Philadelphia means *infidelity;* in Boston, *let us forget.* In Boston, the daffodil means *deceitful hope;* in Philadelphia, it means *chivalry.* Boston says the Roseby willow herb means *celibacy;* Philadelphia does not list it at all.

Another inspired floral approach to education appeared in the presentation of moral tales disguised as "flowers personified." In a large and gaudily illustrated book by this title, translated from the French of a Monsieur Grandville's *Fleurs Animées* and published in New York in 1847, we are given lessons, almost all sad and long, in social behavior. Hand-colored steel engravings are well advertised on the floriferous title page. An elaborate portrait of each beautiful floral heroine clad in her distinctive petals and leaves depicts her appealing innocence. Individual fates, though not portrayed, are universally deplorable and due inevitably to tragic weaknesses of character.

Two tales suffice for examples. In one, Tulipa is beautiful in her little turban and Turkish trousers of leaves and petals. But alas, after being discovered by the sultan and made his favorite sultana, she neglects to improve herself and is cast out by the sultan in favor of Rose Pompom, who is assiduously entertaining and intelligent. Cast out and even away, put into a sack and dropped into the sea by a eunuch. The bag containing her body makes barely a ripple, which soon disappears. The moral: "Beauty without intelligence leaves few traces on the memory."

Another cautionary tale shows Narcissa bending over a stream, admiring her own prettily clad image. Her preoccupation ends in her drowning herself. Conclusion, a curse: "So perish every woman who is without a heart."

To dispel any suspicion that these examples, offered to possible future female botanists, are not to be taken seriously, we append here a review of *The Flowers Personified* in *The Horticulturist* for March 1848. The unsigned review begins: "Here is one of the prettiest fancies, most charmingly carried out, that ever entered the imagination of a writer of tasteful and entertaining books." After three pages of quoted passages, it ends: "To such of our fair readers as do not already possess this ingenious and agreeable work, we beg leave to recommend it as one of the most attractive literary novelties of the season."

These examples of the sentimentalization of flowers and their use as teaching aids may seem to come rather near to each other in times and places of publication in the rapidly expanding United States. But others like them were springing up across the country to attest to the popularity of their subjects—just before the world was set to sober rights and sentiment was divorced from science by Charles Darwin.

High time.

High time, too, for us to move out of the parlor and into the garden.

ANDREW JACKSON DOWNING, leading spirit in the nineteenth-century urge for the beautification of the American home and its surroundings (who will appear later in great depth), ardently believed in the abilities of the American woman. His books were sprinkled with gentle admonitions to "the ladies of the household" to get out into their gardens for their own good as well as that of their gardens. Downing wistfully regrets that his countrywomen are not familiar with the joys he has seen English ladies experiencing, when even a duchess will naturally stoop to pull out a weed. His plans for houses and gardens are often accompanied by what seems no more than gentle reminding.

> There is nothing in the plan of the house or garden that may not be realized by a family living upon a very small income, provided the members are persons of some taste and refinement who appreciate the cheer and pleasure of such a residence sufficiently to take a strong personal interest in it. The master of the premises we shall suppose capable of managing the kitchen garden, the fruit trees, the grass and the whole of the walks, perhaps with the assistance of a common gardener . . . for a day or two. . . . The mistress and her daughters we shall suppose to have sufficient fondness for flowers to be willing and glad to spend three times a week an hour or two in the cool mornings and evenings of summer in the pleasing task of planting, tying to near stakes, picking off decayed

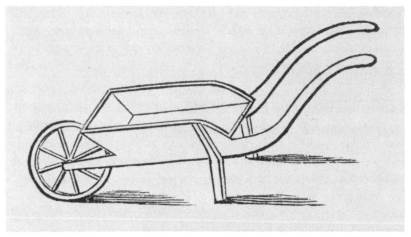

Lady's wheelbarrow, from Mrs. Loudon's *Gardening for Ladies,* edited by
A. J. Downing (New York: John Wiley, 1853).

flowers, and removing weeds from the borders, and all other operations that so limited a garden may require.

In 1843 he edited and published for the first time in America a
book that had previously appeared in three editions in England, Mrs.
Loudon's *Gardening for Ladies and Ladies' Companion to the Flower Garden.* Downing's preface states that he is "confident that this volume will
be a most acceptable one to a large number of persons in this country,"
because in most of the "English works on Horticulture being addressed to those comparatively familiar with everything in the common routine of gardening operations, a considerable degree of previous knowledge of the subject is supposed." With Americans, on the
contrary, he feels that most have to "begin at the beginning for themselves and desire earnestly . . . simple and elementary instructions."
He treads on delicate ground here in recommending Mrs. Loudon's advice as suitable for all beginners, though he wishes it especially to encourage ladies, but he makes his point graciously in the next paragraph.

Mrs. Loudon's works are intended especially for the benefit of
lady gardeners—a class of amateurs which, in England numbers
many and zealous devotees, even among the highest ranks. It is
to be hoped, that the dissemination in this country of works like

the present volume, may increase, among our own fair country-women, the taste for these delightful occupations in the open air, which are so conducive to their own health, and to the beauty and interest of our homes.

Mrs. Loudon's own introduction would disarm any young wife and amateur gardener. Her husband, who will be discussed thoroughly later, was one of the foremost English gardening authorities. "When I married Mr. Loudon," she begins, "it is scarcely possible to imagine any person more ignorant than I was, of everything related to plants and gardening." As we may easily imagine, she says, "I found every-one about me so well acquainted with the subject that I was soon heartily ashamed of my ignorance." Her husband, she says, "of course," was as anxious to teach her as she was to learn.

The result of his instructions, "after ten years experience of their efficacy," she now wishes to make public for the benefit of others. Her very ignorance, she feels, and her having been a "full-grown pupil" herself make her better able to impart her knowledge to others. This she wrote in 1840, so we can see that Downing, in 1843 and three edi-tions later, was quick to see the benefits of her work.

Gardening for Ladies is a small book in itself, of only six chapters: (1) "Stirring the Soil"; (2) "Manuring the Soil and Making Hot beds"; (3) "Sowing seeds—Planting Bulbs and Tubers—Transplanting and Watering"; (4) "Modes of Propagation by Division, viz Taking Off Suckers, Making Layers and Cuttings, Budding, Grafting and Inarch-ing"; (5) Pruning, Training, Protecting from the Frost, and Destroying Insects"; (6) "Window Gardening, and the Management of Plants in Pots in Small Greenhouses." There are a few illustrations of methods and tools, and a sketch of a "Lady's Gauntlet of Strong Leather, in-vented by Miss Perry of Stroud, near Hazlemere."

The second half of the book edited by Mr. Downing is Mrs. Loudon's *Ladies' Companion to the Flower Garden,* published by her in 1842. Mrs. Loudon thought it would be a great convenience to "the possessors of small gardens" to have a "Dictionary of the English and botanic names of the most popular flowers with distinctions for their culture." This is a thoroughly scholarly little volume, not entirely well edited by Mr. Downing in that there is a great deal here of no use to the snow-bound or drought-haunted American garden. However, he has left in

"A Lady's Gauntlet of strong leather, invented by Miss Perry of Stroud, near Hazlemere," from *Gardening for Ladies and Ladies' Companion to the Flower Garden,* by Mrs. Loudon, edited by A.J. Downing (New York: John Wiley, 1853).

some items of absorbing interest to "curious" gardeners in England, like that old favorite of English collectors, Venus's-flytrap, which can be teased with a piece of raw meat. There are some entries of no use to anyone except Mrs. Loudon's neighbors—for instance, "California annuals" are to be held "hardy," which means that, unlike other half-hardy and tender annuals from Africa and elsewhere in the tropics, they can be sown in England in the autumn to bloom the next year.

With two of the greatest nineteenth-century garden authorities coaxing ladies out of doors and into the joys of taking care of their own plants, we might expect success, especially as both approaches are gently persuasive. But such tactful wheels ground slowly. In the nineteenth century American ladies do seem to have been reluctant to engage in outdoor exercise beyond such social graces as strolling or a game of croquet. With sentiment and delicacy in the ascendant, fainting appeared a more suitable and graceful expression of female power than seizing a spade and rushing out to dig. Florence Nightingale ruled from her couch nationally to a degree she might never have attained on her feet. A Victorian lady could show her disapproval almost to order by taking a deep breath inside her tight stays, then holding it until her circulation stopped and she keeled over, looking her very best. With such a weapon fashionable, why should the American female seek to excel with a hoe, even a little hoe, or a tiny wheelbarrow made especially for her? Even if Englishwomen chose to pull weeds.

But the ladies were not to be spared. Stronger tactics were adopted by other writers. A sudden spate of small books insisted on the extreme importance of floriculture, especially in relation to the happiness of the family and the improvement of the home landscape. One of these authors, John T. C. Clark, put it very clearly in his book, *The Amateurs' Guide and Flower-Garden Directory,* published in Washington in 1856. In his preface, he states:

> The love of Nature affords us the purest delight and is implanted in the human breast. . . . Nature is most lavish in her gifts, and in order to appreciate and enjoy them, we should listen to her voice and study well her teachings, for they will surely inculcate a tone of refinement, afford pleasant and healthy employment, and give us exalted views of her Creator.

Whereupon, like all the other authorities, endeavoring (on the "repeated demands of kind friends") to portray the "practical study of Floriculture," he addresses himself to the "wants of the Amateur." He hopes to persuade others to "make the acquaintance of Flora, whose flowery paths abound with innocent pleasures."

We recognize him as both honorable and reliable, because he acknowledges that for the "generic or specific names of many plants" he is indebted to, "among other standard works," John Claudius Loudon's *Encyclopaedia of Plants.*

In his introduction, Clark feels all nature rejoicing with him, and "the swellings of his own bosom are but the vibrations of that all-pervading system of harmony that thrills throughout that vast extent of Creation." In fact, because

Nature is most industrious in adorning her domains. . . . Man, to whom this bounty is addressed . . . should adorn his home—the dwelling of his wife and children—with pleasant objects, and with all those attractions which will make it cheerful. What will tend more to this end than a flower garden, filled with beautiful flowers, imparting their fragrance, elevating and purifying the soul of the beholder? If this is done, the home will become the abode of cheerfulness. . . .

Where flowers are planted, a home becomes a tasteful residence, while its intrinsic value is greatly enhanced. Cultivated taste gives beauty and value to property, and the small cost of a flower-garden, so far from being a useless expense, as some assert, adds to the money-value of the property. . . .

Children learn to love every flower. . . . In after life their affections will cling to the beautiful and hallowed spot where first they beheld the beauties of Nature . . . whose lessons were imprinted on their young hearts by a fond and affectionate mother. The eyes of the father become opened to the influence of the lessons which Nature imparts, and he fondly cherishes every plant which has afforded so much pleasure to his children and comfort to his household.

And now he pounces, although gracefully: "The author would respectfully appeal to all to encourage the study of practical Floriculture, and particularly the LADIES."

So many English writers on landscape gardens had urged "consulting the Genius of the Place" that credits for the origin of this expression vary widely in nineteenth-century American garden books. The thought is correctly attributed to a much earlier authority, Virgil, whose principles still held sway in much gardening and farming lore. Gallantly, Clark contributes his own version of the gender of this genius.

Proceeding to instruct his lady amateurs on "Walks and Beds," he admits to a "great diversity of opinion in regard to the rule or plan on which grounds should be laid-off." He personifies the "genius," handles the controversy deftly, and sums up several contemporary styles in three notable sentences.

Page from *Farming for Ladies* (London: John Murray, 1844), author unknown, even to the Murrays. "Her Majesty's poultry-yard at Windsor is situated in a small pleasure-garden . . . the henhouse . . . a simple though fancifully decorated cottage, displaying considerable taste in the architect."

Some contend for straight or parallel walks and beds, while others maintain that the whole beauty of the garden is marred unless it be cut up into serpentine walks and irregular or fancy-shaped flower-beds. And others again affirm that in the blending of the plants and shrubs together, so as to hide all artificial or studied effect, consists the acme of perfection. Where such a difference and taste exist, we would not willingly become the umpire but as every person will be governed in part, at least, by their own opinion and taste in these particulars, we would venture to suggest that, in planning and laying out grounds, if they will consult "The Genius of the Place" and act in accordance with such suggestions as she may point out, they no doubt will be able to effect this object to the mutual satisfaction of themselves and their critical friends.

Altogether, we can be indebted to this honest little handbook for its picture of garden times in the mid-nineteenth century. Its lists are divided into designatedly tender or hardy annuals, biennials, and perennials. Botanical names are followed by the common, and he adds dates of blooming, colors, and habits. He lists also "Bulbous and Tuberooted Flowers," "Deciduous Plants" (which include shrubs and vines), "Evergreens," and "Roses." He advises that walks should be at least two and a half feet wide and well drained. He discloses that box, "the dwarf variety (*Buxus suffruticosa*)," is generally employed, "if kept neatly trimmed and regular," to add to the beauty of the garden "especially in winter." We have not seen box much employed until the nineteenth century, although it is now used extensively in restorations of much earlier gardens.

The idea that there is nothing women could not, suitably encouraged, accomplish in their gardens must have given some of the more thoughtful ladies pause.

BUT, obviously, the ladies are not to be given time to stop and think. Books directed toward them seem to spring from the shelves. In 1856 another small book appeared, in New York, called *The Requirements of American Village Homes, Considered and Suggested with Designs for Such Houses of Moderate Cost.* Three men of determination and self-confidence, Henry W. Cleaveland, William Backus, and Samuel D.

Backus, obviously partners in building "homes," had combined to tackle the problem of producing "pleasing" architecture and gardens on the lowest economic level yet considered.

After fifteen chapters of exhortation and practical advice (on the influence of a home on its occupants, the value of permanence, siting, one-story cottages versus two-, and tasteful interiors), the last two chapters are reserved for the improvement of grounds and the garden. "Improvement" includes drainage, with clever and sanitary ways to save waste water by use of a stench trap, and designs for relatively indestructible picket fencing.

And then comes a really original inspiration.

"Cleaveland and Backus Brothers," as they are designated on the notice of being "entered according to Act of Congress in the year 1855," give over the last chapter on flower-bed designs and lists of plants to Peter B. Mead, Esq., secretary of the New York Horticultural Society, who is more than ready for the ladies. By comparison, the flowery Mr. Clark's expectations of "LADIES" seem modest. Mr. Clark sought honestly to inform from ground level up—in cases of horse manure, skills of grafting and layering, even varieties of tools. Mr. Mead's ambition for the ladies knows no such humble bounds. To be extra helpful he has drawn up two possible flower beds and given explicit directions for one.

It is laid out in a square, thirty-six feet by thirty-six, with a center of five conjoining circles surrounded by paths. The outer halves of four larger circles contain planted beds of semicircular shape, which are in turn surrounded by paths connected to each corner's quarter-circle planted bed. The whole forms a nest of curves. But this design, according to Mr. Mead, is "easily formed. Two sharpened sticks connected by a string, are the only instruments required." He points out that a "selection of plants proper for each spot and so arranged with reference to size, colors, etc., as to produce a pleasing and harmonious effect will be found in the following instructions."

We will begin with Mead in the center, with a shrub, *Spiraea reevesiana,* surrounded by four roses in the four circles: Geant des Battailles, Caroline de Sansal, Prince Albert, and Pius IX. In each of these beds as well, equidistant from each other and about a foot from the edge, we are to plant three of the following: *Spiraea filipendula, Plumbago larpen-*

tiae, Anemone japonica, Crucianella stylosa, Myosotis palustris, Hepatica triloba, Dodecatheon meadia, Alyssum saxatile, Convallaria majalis, Aquilegia glandulosa, Sedum sieboldii, and *Chelone barbata.*

Any lady who had done a bit of gardening before Mr. Mead caught her eye would know that the varying heights and sizes of these suggested blooms will take a lot of planning to squeeze into their allotted spaces. But Mr. Mead harbors no fears.

The four outer half-circles are all to be planted alike, with the following plants. In the center of each semicircle go four roses—Hermosa, Mrs. Bosanquet, La Reine, and August Michelet. In the corners, or the horns of the semicircles, are phloxes. At the centers of the back curves, antirrhinums are planted next to delphiniums and *Lychnis chalcedonica, Dictamnus rubra, Penstemon gentianoides,* and *Campanula grandiflora.* The inner borders are to be filled with *Valeriana rubra, Oenothera fraserii, Lychnis viscaria, Veronica spicata, Penstemon atropurpureum, Lupinus polyphyllus, Aconitum napellus,* and *Aconitum speciosum.* In the middle of the inner curves we are to place *Dracocephalum speciosum, Valeriana officinalis, Spiraea lobata,* and *Spiraea americana.*

Mr. Mead continues. In the four great "outer borders" we are to have "a good collection of chrysanthemums, clumps of tulips, hyacinths, narcissus, jonquils and crocuses." And "bedding plants" are "indispensable." Among the best, he says, are verbena, petunias, cupheas, scarlet geraniums, nierembergias, gaillardias, and so forth. And we are not to forget "Dahlias, Gladioluses, etc."

One wonders if any eager ladies-cum-amateurs were inspired by this erudite gentleman. One hopes he may later have been forced to have a try at taking his own advice. We share his dreams here to show what ladies may have been up against if they strayed from the gentle encouragements of Mr. Downing and the practical experience of the indomitable Mrs. Loudon.

Far easier to draw a deep breath, hold it, and faint away.

NOTE TO CHAPTER 6

1   John Parkinson, in his great book on gardening and plants, *Paradisi in Sole, Paradisus Terrestris,* published in 1629, said he included "Buglosse and Borage" in his "Garden of Pleasant Flowers," rather than in a kitchen garden, because "anciently . . . their

flowers having been in some respect in that they have always been interposed among the flowers of women's needle-work." And, in considering the jonquil, he warned that one must never place a bowl of them in a room where ladies were likely to sit. There his concessions to women ceased. Incidentally, when we find that a hundred years later the jonquil stood for "lust," and that in another hundred years its message had been modified to "desire to make your acquaintance," we can realize the pitfalls that await those endowing plants with other than horticultural characteristics.

SEVEN

Societies and Shows

ITH the coming of an industrial revolution in England, a revolution in domestic horticulture also arrived. The onus—as well as the honor—of promoting the propagation of new flowers descended from being the province of a few rich individuals and their expert gardeners to becoming a backyard hobby of the so-called working classes. "Mistress Tuggy's" sixteenth-century carnations, immortalized by Gerard, led the way for a proliferation in the nineteenth century of innumerable Tuggyish varieties and oddities in carnations and in auriculas, especially.

In the north of England, men home from work earlier than ever before—and with more time and money to spend—bent over their specialties in little glass-covered cold frames, squeezed into brick areas beside washing arrangements, and readied them for shows. Competition had become general. Although horticultural philanthropists financing distant expeditions for botanical research still hoped to be the first with the best, these lesser mortals in industrial cities, equally keen, kept their secrets hidden until they could expose them to envious gaze at local flower shows. That these shows were often, by courtesy of the large landholders, held on vast lawns under tents—one for the gentry, another for all else—did not matter when it came to showing. There everyone, under another great spread of canvas, competed on equal levels, and merit alone won the prize.

In nineteenth-century America, with the country expanding westward like a great tide, there was less control over matters horticultural,
less organization, less opportunity to see what was going on elsewhere. Little groups gathered in seedsmen's offices (often in the most
important buildings in the business sections of large cities) became
pivotal to the spread of horticultural news. They organized themselves
into societies for the promotion of their interests, and they took to the
printed word, in catalogues, books, and newspapers. They apparently
buried business rivalries beneath a common goal. Several seedsmen
with offices in the same building could see the advantage in creating
for themselves and their customers a horticultural society quartered in
that same building. Subscribers, members, borrowers and lenders of
books, and givers of advice were welcomed. Board members were
chosen. As the societies burgeoned, there had to be occasions for exhibiting the rarest and most spectacular results of the members' efforts.

The hand-in-glove relationship between the early horticultural societies and the seedsmen's offices is illustrated in the frontispiece of
Washburn and Company's 150-page, hard-cover catalogue, published
in Boston in 1869. In the lovely picture, an elaborately dressed lady,
with a rake, sets out roses from pots. Elegantly clad children help her,
while a ruffle-capped nurse holds a baby. An elderly lady in the background is training a vine by the door. Nothing could more beguile the
amateur than this inspirational scene, called simply "Gardening." The
booklet itself is entitled *Amateur Cultivator's Guide to the Flower and
Kitchen Garden: Containing a Descriptive List of Two Thousand Varieties of
Flower and Vegetable Seeds; also a List of French Hybrid Gladiolus.* There
is a splendid picture of Boston's new Horticultural Hall set in here, and
the text continues beneath that splendid facade ("raised and imported
by Washburn and Company, Seed Merchants, Horticultural Hall, No.
100, Tremont Street"). So we see this building contains their offices, as
well as those of the Massachusetts Horticultural Society. A slight confusion in the dedication—"To our amateur friends and customers"—
cannot detract from the feeling that for all of us, with this booklet in
hand, all things are possible.

Dr. William Darlington describes a typical setting for those early,
impromptu gatherings of the horticulturally inclined, sketching the atmosphere of Bernard M'Mahon's shop, in the 1857 preface to the elev-
enth edition of M'Mahon's *American Gardener's Calendar.*

1869.   ESTABLISHED 1845.   1869.

WASHBURN & CO.'S

AMATEUR CULTIVATOR'S GUIDE

TO THE

Flower and Kitchen Garden:

CONTAINING A DESCRIPTIVE LIST OF

TWO THOUSAND VARIETIES

OF

FLOWER AND VEGETABLE SEEDS;

ALSO A LIST OF

FRENCH HYBRID GLADIOLUS.

HORTICULTURAL HALL.

RAISED AND IMPORTED BY

WASHBURN AND COMPANY,

SEED MERCHANTS,

HORTICULTURAL HALL, No. 100, TREMONT STREET,

BOSTON, MASS.

Title page, Washburn & Co., *Amateur Cultivator's Guide;* the frontispiece
for this catalogue is found in the Preview to this book.

Many must still be alive who recollect its bulk window, ornamented with tulip-glasses, a large pumpkin, and a basket or two
of bulbous roots; behind the counter officiated Mrs. M'Mahon
with some considerable Irish accent, but a most amiable and excellent disposition, and withal an able saleswoman. Mr. M'Mahon
was also much in the store, putting up seeds for transmission to
all parts of this country and Europe, writing his book, or attending to his correspondence, and in one corner was a shelf containing a few botanical or gardening books, for which there was then
a very small demand; another contained a few garden imple-
ments, such as knives and trimming scissors; a barrel of peas, and
a bag of seedling potatoes, an onion receptacle, a few chairs, and
the room partly lined with drawers containing seeds, constituted
the apparent stock in trade of what was one of the greatest seed
stores then known in the Union, and where was transacted a considerable business for that day. Such a store would naturally attract
the botanist as well as the gardener, and it was the frequent lounge
of both classes, who ever found in the proprietors ready listeners
as well as conversers; in the latter particular they were rather remarkable, and here you would see Nuttall, Baldwin, Darlington,
and other scientific men, who sought information or were ready
to impart it. Mr. M'Mahon was esteemed by these, and in several
botanical works his knowledge is spoken of with great respect and
consideration.[1]

The gentlemen gathered in M'Mahon's shop represented the first
spark of the Pennsylvania Horticultural Society. A similar group in another seedsman's shop in Boston, at about the same time, generated the
Massachusetts Horticultural Society. We will return to these two societies in depth in a moment, because they serve as examples of the
burgeoning of groups for the promotion of horticulture in nineteenth-
century America, and because each is still in successful operation, carrying on the original purposes of its founders.
The model for all these budding organizations was the Horticultural
Society of London, organized in 1804 and chartered in 1809. To its di-
rectors through all the years of its existence the whole world owes a
debt. Today, in the twentieth century, up and down and across the
United States and into Canada, similar groups promote the well-being

of the countryside. Horticultural schools have proliferated; landscape gardening courses have penetrated polite academic circles. Garden clubs and their state federations flourish wherever plants are grown. And these all began with groups of rather portly, bearded gentlemen in frock coats, in the early part of the nineteenth century, seated in large and comfortable chairs in the offices of one or another of the seedsmen and nurserymen.

Not all the societies held on. The New York Horticultural Society was organized in 1818 and incorporated in 1822 with high ambitions. There was to be a ten- to twenty-acre garden for the promotion of horticulture and the study of botany, particularly for the cultivation of fruit trees. A lecture hall, a library, and a "botanical cabinet" would assist in the added inducement of a professorship of botany and horticultural studies. In 1823 one Dr. Hosack, elected a member of the society in 1822, tried to get the society to take over his originally flourishing, though then neglected, botanic garden, but he was disappointed. However, in 1824 he was made president of the society. André Parmentier, our first professional landscape gardener, of whom we shall see more later, also became a member. But John Torrey may have been a prophet when he complained, in 1827, while teaching botany at the College of Physicians and Surgeons, that the area lacked interest in the natural sciences. For, within fifteen years of its founding, the first attempt to start a horticultural society in New York failed. Similar societies begun in Geneva and Albany were also short-lived.

Our stalwarts, the Massachusetts and Pennsylvania societies, often referred to themselves as "sisters," which makes reviews of their meetings the livelier, because they were so obviously in constant competition. For example, when the Pennsylvania society received two bottles of white currant wine from John Prince of Jamaica Plain near Boston, to show how their report that currant wine did not improve with age was incorrect, the society graciously replied that they had found the wine slightly acid and not equal to some of their recently exhibited samples, but they unanimously presented their thanks to Mr. Prince. In retrospect, Boston seems over the years to have excelled in fruits and Philadelphia in flowers, but the balances were drawn finely.

An account of the initial day of the Massachusetts Horticultural Society includes many of our friends from this study and gives a fair picture of these worthy gentlemen. The office of the magazine *The New*

England Farmer had been established in 1827 at 52 North Market Street, with Thomas Green Fessenden as editor. On the ground floor, John B. Russell, its owner, maintained a seed store, which had become a gathering place for those interested in either enterprise. Judge Buel of Albany had suggested to Russell that a horticultural society should be started in Boston, and when an insurance man of Dorchester, Zebedee Cook, Jr., wrote an article for *The New England Farmer* calling for a similar organization, Fessenden echoed their sentiments in an editorial. The aged John Lowell also supported the idea. In the February 20 issue of *The New England Farmer,* "those interested" were requested to meet at Zebedee Cook's office on the twenty-fourth. On that day at noon, in the middle of a howling blizzard, sixteen men appeared, including John Lowell, wrapped in blankets and brought by his neighbor, Cheever Newhall, in his sleigh.

The purpose of the meeting was expressed by the future society's first president, Gen. H. A. S. Dearborn.

> It became apparent that a zealous cooperation of all persons interested in gardening was required for producing a more general and speedy extension of scientific and practical knowledge in all its branches; and in the winter of 1829 a number of gentlemen of Boston and the adjacent towns determined to attempt the establishment of a Horticultural Society.

The Pennsylvania Horticultural Society was organized earlier and in better weather than its Boston sister, in 1827, with fifty-three dues-paying members. In 1828, David Landreth was authorized to order copies of Loudon's *Gardener's Magazine* from the first number on to a continuing subscription. By 1829, sixty-five more members were nominated, including three ladies.

At monthly meetings members were expected to bring any particularly beautiful or unusual flowers, plants, or fruits they had grown, and in November of 1828 members exhibited forty specimens of plants and flowers, including fifteen varieties of pears and apples, some American grape wine, and some fine cauliflowers and broccoli. "Wine was frequently received for test and comment." The first public exhibition was held in June 1829, probably the first important flower show in America. By 1832 the show became an annual affair, but relocated to

September to allow for the showing of fruits. Introductions of the tropical and semitropical plants then becoming popular were made by our nurserymen friends Landreth, Buist, Hibbert, and McArann. The most popular plants were the camellia, the rose, and the dahlia.

From their accounts of their meetings, we learn that Mr. Landreth "brought forward two *Azalea indica,* natives of China," and Mr. Hibbert reported his Chinese tree peony in bloom. And, fortuitously enough, "Our country has been recently pronounced in a foreign journal to be rich beyond all others in stores of botanical wealth." This, in the magazine to which they had already subscribed handsomely. In all fairness, the society admitted to a soil and climate in Philadelphia "adapted to a great variety of vegetable products" and a community "congenial to the pure and primitive employment which consists in the cultivation of them." It is "to inspire this zeal, to multiply the sources of information and bring them within reach," that the Pennsylvania Horticultural Society had been formed.

Here was where the poinsettia was first exhibited, in the company of magnolias, tree peonies, a double white pomegranate, pelargoniums and carnations of many varieties, and a surprise from the Cape of Good Hope, *Strelitzia reginae.* And in no time in Philadelphia they were propagating camellias, stock gillyflowers, and double primulas. Another exhibition faithfully reported in the news contained the "Greville rose" (our "Seven Sisters," the only Chinese species of rose hardy enough for New England), and the Petre pear, sent to John Bartram by one of his most faithful subscribers, young Lord Petre, beloved by Collinson. Six roots of the *Colchicum autumnale* were received from London, with a notice of success with it in treating gout. And within ten years, in 1881, Caleb Cope, assisted by his gardener, Thomas Meehan, succeeded in getting a bloom on the greatest lily in the world, *Victoria regia.* The seeds were sent him in a bottle by William Hooker. The list of plants exhibited is dazzling. Within another ten years the Buist roses had become famous and Mr. Buist had named a new *Azalea coronata* "Caleb Cope."

By their seventeenth annual exhibition, the society's exhibitors were having troubles with the design section. Where they had prided themselves on the "chaste and novel" arrangements of their early shows, things had begun to get out of hand. No prizes were awarded for this

"Artificial Rockwork. Constructed by G. M. Kern and exhibited at the Fall Exhibition of The Cincinnati Horticultural Society—1854." Frontispiece for *Practical Landscape Gardening* by G. M. Kern (Cincinnati: Moore, Wilstach, Keys, 1855). The modern "rock garden," simulating a natural habitat for alpine plants, arose from fantasies of "rockwork," as seen in this exhibition for a horticultural show and the picture of Mrs. Lawrence's rock archway topped by a winged figure and balanced by matching fountains (p. 286).

year's offering, ostensibly because they were "so few." (A pity, they admitted, because they had cost so much in time and money.) Archibald Henderson's ten-foot flower stand had a pool at the bottom, complete with fish, and ended at the top in a stuffed bird, above a succession of baskets, each raised by four arms. The baskets and cage were all made of gomphrena flowers. In another entry, a sofa and chairs of a rustic furniture set were made of greenery and the words "Sit down" were picked out in white gomphrenas on the sofa seat.

We now turn our attention to the sister society, at the fourteenth anniversary of the Massachusetts Horticultural Society in 1842, when ladies were admitted to the dinner for the first time. Quite a debate had preceded their inclusion, as the choice appeared to fall between them and wine. In the end the ladies won, and at the dinner—attended by 200—the Reverend Mr. Winslow was able to congratulate President Wilder on "the vast improvement he had made in our public festival in exchanging the intoxicating cup for the more elevating and refining gratification realized by the presence and smiles of women." The hall was decorated with a huge bouquet from floor to ceiling at either end, one topped by a banner reading "Fourteenth Anniversary of the Massachusetts Horticultural Society" and the other by a tablet: "The world was sad—the garden was a wild. / And man the hermit sighed—till woman smiled."

By the time the Seventeenth Annual Exhibition was held, in Faneuil Hall, the Boston society had a new building. The dinner was attended by 600. Exhibitions had been so great as to tax the accommodations— one exhibitor alone sent 240 varieties of pears—and there was an added competition in "designs." Besides a Greek temple, a Chinese temple, and a Gothic pyramid, all worked out in greens, asters, and amaranths, there were a harp of evergreen with strings of wintergreen and arborvitae, a lyre with suspended grapes, a spread eagle of asters with a chain of rose-hip beads hanging from its beak, a plow made of asters, and a Newfoundland dog—executed in pressed "black" hollyhocks and grayish moss—carrying a basket of flowers (this last by a Miss Russell, who received a premium of six dollars).

We must realize that at this time, when oratory flourished like a green bay tree, the guests were in for a treat. President Wilder alluded to a meeting of sixteen years before when there had been less than a hundred baskets of fruit and $200 in premiums, whereas this occasion offered 1,400 dishes of fruits with premiums of over $1,300. There were formal toasts—so something had changed in the last three years—and Edward Everett, reputedly the leading orator of his time, was introduced, just arrived that morning from serving as ambassador at the Court of St. James. He made a short but charming speech expressing a preference, after his recent rough crossing, for listening to others. Daniel Webster, "our Marshfield Farmer—all head in council—all wise in speech," rose to note with gratification the growth of the society he had helped to form. He remarked upon its being our fortune in New England to live beneath a somewhat rugged sky and till a somewhat hard and unyielding earth but remarked that hardness seems necessary "to excite human genius, labor and skill." He greatly doubted, however, that "all the luxuriance of the tropics and all that grows beneath the fervid sky of the equator can equal the exhibition of flowers made today amid these northern latitudes." Then he, too, complimented the ladies by saying, "the plants of the garden are cultivated with us by hands as delicate as their own tendrils, viewed by countenances as spotless and pure as their own petals, and watched by eyes as brilliant and full of lustre as their own beautiful exhibitions of splendor." Josiah Quincy, president of Harvard, remarked upon the blessings of well-directed industry; Caleb Cushing paid yet one more tribute to "woman—our equal—shall I not say our moral superior"; and someone toasted "Our Merchant Princes," one of whom had sent $1,000 for premiums "for producing Trees good for food and Flowers pleasant to the sight." An added adornment to the meeting was the presence of Mrs. Hamilton, widow of Alexander Hamilton and daughter of Gen. Philip Schuyler. Her health was proposed by Daniel Webster, who claimed that "neither great name would be forgotten while devoted revolutionary services shall be remembered." Witty speeches and songs followed. A hymn was sung. More toasts. And Charles Mason Hovey, editor of the *Magazine of Horticulture,* praised famous men: John Lowell for the spread of information upon every branch of horticulture and for his importations from England of many of our finest fruits; Judge Buel for his *Cultivator;* Fessenden for his *New England Farmer;* and above all Loudon, "the Walter Scott of horticultural literature." More toasts, and the brilliant festival ended.

Prosperous years sped by for the society, marked by ever more glamorous exhibitions and memorable events, such as the introduction of Ephraim Bull's Concord grape and Dana Hovey's seedling pear. Mount Auburn Cemetery, an early project of the society, which was so important that it will be covered separately, continued to yield a fine income, which was set aside toward the building of a suitable hall for the society. The *Victoria regia* lily was raised in Salem. At Gore Place there was word of a machine, newly invented in England, for close-cutting, gathering, and rolling lawn grass all at the same time. A "Committee on Ornamental Gardening" visited Mr. Fay's garden in Chelsea and found Mrs. Fay "arranging, planting and cultivating with her own hand the flower treasures of the earth" (subsequent lamentation that "so few of the ladies of our land imitate her example").

By 1862, with the war at its height and economy the watchword, exhibitions were small, but "improved" hybrid perpetual roses were at their peak, as were dahlias. *Lilium auratum* was introduced. Mr. Breck recommended weather reporting. Fruit had never been more abundant. In fact, fruit prices were so lowered that pears were given away on State Street to the newsboys. At a meeting in 1877 on sterilization and cross-fertilization, a crisis over Darwin and his publications was tactfully resolved by Mr. C. M. Hovey's admission that, with all the good Darwin had done, his writings had been "to some extent sensational." (In 1882, however, a united tribute to his memory was requested by the society and written by Asa Gray.)

In 1878 the Flower Committee, at the urging of three clergymen, undertook to benefit the children of the "laboring classes." Three ladies—working through the Sunday schools of various churches—began to instruct them in horticulture through the planting of window boxes. This original effort to bring a feeling for beauty into otherwise dreary homes gathered such strength that the ladies requested funds from the society to further their enterprise, and children's horticultural shows were organized with great success.

A review of this period, made for the Boston society by unidentified individuals, is helpful to gardeners studying the plants and plans of the times. The reviewers were

> struck with the great progress in raising improved varieties, particularly in flowers. Not only was the improvement of the old favorites—the gladiolus, camellia, Japan lily, phlox, rhododendron, and petunia—continued, but the carnation, paeony, delphinium,

pelargonium, coleus, amaryllis, polyanthus, verbena, dracaena, cyclamen, pyrethrum, and other flowers, were made subjects for the florist's art. Moreover, this improvement was pursued in a more systematic and scientific manner than before, hybridization being more generally practiced. Probably this activity was in part due to the war, which exercised an unfavorable influence on the exhibitions by increasing the cost of importing novelties from Europe, caused our florists to rely more on their own exertions for the production of improved varieties, and thus led to more desirable results.

The reviewers were impressed by the richness of the 1870s in American horticulture and by

the great number of rare, curious, and beautiful plants introduced. The wealth of our greenhouses and hothouses in this respect was revealed by the exhibition of 1873, when Music Hall was wholly filled by the most beautiful display of plants and flowers that had ever been made on this continent. The Orchidaceae were more largely represented than ever before.

Growers were ready to accept new plants and new trends in gardening.

The taste for agaves, cacti, sempervivums, and other succulent plants, grew up during this period. Equal activity was shown in the introduction of hardy plants for the ornamenting of our gardens; and the *Deutzia crenata,* the *Viburnum plicatum,* the *Hydrangea paniculata grandiflora,* the new hardy varieties of the clematis, and the *Aquilegia chrysantha,* novelties of this period, are destined to find a place in every garden: indeed the beautiful hybrid varieties of the clematis are one of the triumphs of horticultural skill; while the almost innumerable forms of the aquilegia, discovered or originated, are but an instance of the improvement in flowers formerly known to us by a few types, or only a single one. The many and beautiful new conifers exhibited have attracted much attention, and, if but a tithe of them prove adapted to our climate, they will be most valuable additions to the beauties of our lawns. The rhododendron show was not only an important event in the history of the Society's exhibitions, but will doubtless form an epoch in the cultivation of these beautiful plants.

The members' interest in native flowers,

> which, in the early days of the Society, was intermittent in charac-
> ter, in this period, became, through the labors of several zealous
> collectors, not only more extensive, but continuous; and the deli-
> cate forms of ferns were added to the flowering plants exhibited.
> Though hitherto a comparatively small number have been intro-
> duced to our gardens, there are now indications of a more general
> attempt at the cultivation of native plants.

Best of all, attention was paid to

> the bedding system, and to carpet and ribbon gardening, as men-
> tioned in the reports of the Garden Committee . . . styles of gar-
> dening introduced at about the commencement of this period.
> Not far from the same time we became acquainted with the
> iresine, the coleus, and the alternanthera, without which the sys-
> tem could never have been carried to the extent which it has at-
> tained. Along with these and older plants, which were propagated
> in immense numbers for this purpose, came the infinite variety of
> pelargoniums, their production being stimulated by the bedding
> system, in which ribbon and carpet gardening produced effects
> more brilliant than had previously been seen in our gardens. Sub-
> tropical gardening, which is also mentioned in the reports of the
> Garden Committee, was introduced somewhat later than carpet
> and ribbon gardening, and with its groups of cannas, caladiums,
> dracaenas, tritomas, wigandias, etc., gave new aspect to our gar-
> dens, while isolated specimens of palms, bananas, tree ferns, and
> similar exotic plants, produced on our lawns an air of refinement
> and distinction previously unknown. These, and the multitude of
> ornamental foliaged plants, both hardy and tender, which now
> enrich our gardens, form the most characteristic feature of the
> present era in horticulture.

But by this time there was also a reassessment of public taste and a
retreat from the excesses of the past. A florist in Boston reached his
limit when he was asked to furnish a bullock's head in flowers for the
funeral of a butcher, and the ladies of the Flower Commitee, hearing
of this, joined with him and publicly protested the floral "mess" in

local cemeteries. In Philadelphia we have seen the fading of the competitions in extravaganzas of floral art.

The libraries of the two societies burgeoned into national importance. The library of the Massachusetts society, from being, as it was reported proudly, "fifth in the world," grew to be first. With a fund dedicated to the buying of books, the society had begun by subscribing to Audubon and gone on to collect the great botanical works of the world as they were written and published. Certain nonreaders objected to this use of funds. In 1885, in answer to the complaint that "useless" books were being bought—as evidenced by the modest circulation of some—Robert Manning, eminent pomologist, announced that

> as regards a library whose maintenance insures . . . the preservation of the best fruits of advanced research, in a repository accessible to scholars and students . . . the benefit reaped from it by the community cannot be reckoned by any method of statistics. . . . Decidedly the most valuable part of our library consists of books by no means adapted or intended for general circulation.

Unhappily, 100 years later, with the library enormously increased in value as well as interest and importance, the circulation counters won, and the library was destroyed as a national treasure, for benefits yet unproven.

And so they continue, our societies, still great powers, housed in imposing buildings, loyally supported by their seedsmen and nurserymen, merchant princes, ladies' committees, and no doubt the grateful offspring of the laboring classes who benefited so long ago from instruction in the value of window boxes.

NOTE TO CHAPTER 7

1 It is more than fair for Dr. Darlington to include himself in this gathering. For much of the nineteenth century until his death in 1863 he was, as Joseph Ewan described him, a "physician, botanist, friend of botanists, author, correspondent and antiquarian. He gathered historical facts on the progress of botanical science, which in the nineteeth century amounted to plant collecting, description and writing of florulas." Darlington's *Memorials of John Bartram and Humphry Marshal* is invaluable in studying the eighteenth century. His consideration of the sensitivies of others serves here to make a point about

the nineteenth century. In 1837, John Bartram had detailed what was certainly the first scientific exercise in hybridizing in this country. Darlington edited carefully for his mid-century readers, who were having a hard time assimilating Darwin. Describing Bartram's experiments, Darlington left out so many words and details he felt might offend his nineteenth-century readers that the experiments sound more like divine intervention than the practical manipulation of the flowers' sexual organs.

Laying on of Hands

EIGHT

Early Links

SEARCHING for influences can be a dangerous procedure, one that involves leaping across gulfs, grasping at dangling ends, and setting too much store by one's own intuitions and associations. Historical events, politics, and philosophies leave their imprints. One can attribute too much to too many, or not enough to one individual. We are on safer ground observing recorded plantings, even of ideas. Where these ideas actually originated may be beyond our careful digging; but with others' notes in our hands, with others' careful acknowledgment of indebtedness for this or that, we can construct a living chain of individuals that will carry us through the whole nineteenth century in American landscape gardening.

Running through the century under consideration, we can see a strong pattern, a chain of individuals, each knowing the one just before and the one just after himself, handing on the new concept of landscape gardening for the good of the New World. The chain begins with Dr. David Hosack and continues through Parmentier, Downing, Vaux, Olmsted, and Eliot. The living communication of ideas about landscape gardening in North America can rest with these six individuals, although many others contributed their efforts and will be considered along the way.[1] We may find it difficult today to understand the extent of their dedication, the urgency of the need for their ideas.

For even though plants for medicine and food crops continued to be of vital importance, landscape gardening became second to none in the shaping of the new land and its growing cities. Gardening, which had been practical in the seventeenth century, focused on "meate and medicine," blossomed into being "for use and delight" in the eighteenth, and in the nineteenth became part of the framework for a new country.

DURING the previous centuries, the concept of gardening for beauty and national benefit did not have much urgency, although William Penn had planned for parks and Benjamin Franklin had disclosed the importance of street-tree planting to public health. At the beginning of the nineteenth century, Dr. David Hosack began to promote the use of gardening for the general good and for the enhancement of domestic surroundings.

A vibrant character—with the most dynamic personality in our succession—Dr. Hosack was well known to many and well meaning toward even more. He became an important figure in the new democracy. A New York civic leader, a generous host at his Hyde Park estate on the Hudson, a born innovator, he handed on his many blessings like any good Puritan.

David Hosack was one of the phenomena attendant upon the American Revolution. Like George Washington and many others less known, he was born into loyal and royal affiliations, changed his allegiance, suffered no resentments or retaliations, and retained friends abroad. He even pursued his higher studies and more glamorous hobbies with foreign support. In the end, with the forming of the United States, he (like Washington) devoted all his talents and much of his fortune to bettering the affairs of the new country.[2] Most important for us in the history of American gardens, he started the first American botanical garden and created upon the bank of the Hudson one of the foremost garden estates.

Born in 1769 in New York City to a father who was a British artillery officer from Scotland, David went to school in British-occupied New York and then went on to study medicine in both New York and Philadelphia. Leaving his wife and child in the care of his parents, he went to study more medicine, as well as botany and mineralogy, in the British Isles. Well aware that his Philadelphia preceptors, Benjamin

Smith Barton and Benjamin Rush, considered botany "a necessary part of education," Hosack sought instruction from James Edward Smith of the Linnaean Society and, while in London, made daily visits to the botanic garden of William Curtis.

Hosack arrived back in New York in 1794, his reputation newly enhanced by a successful handling of an outbreak of typhus among the steerage passengers. Within a short time, he acquired equal fame for the treatment of a yellow fever epidemic in New York. He went into partnership with his former teacher, Samuel Bard, and began to teach at Columbia College.[3]

One of Hosack's dreams involved the founding of a "small garden" where he could grow "such plants as furnish the most valuable medicines and are most necessary and valuable for medical instruction." He hoped that New York would "have the honor of beginning the establishment of this kind of garden in this country." He was not the first to suggest such an enterprise. His predecessor in the Columbia teaching position, Dr. Samuel Mitchill, another New York physician who had studied in Edinburgh, had requested funding for a similar project from the state legislature in 1794, arguing that a

> garden is nearly connected with a professorship of botany. Lectures must always be lame and defective without a garden. . . . A botanical garden is not only the most useful and most important of public improvements, but it also comprises within a small compass the history of the vegetable species of our own country: and by the introduction of exotics, makes us acquainted with the plants of the most distant parts of the earth. . . . Likewise, by facilitating experiments upon plants [it is] one of the means of affording substantial help to the labours of the Agricultural Society.

In 1800 Hosack himself petitioned the state legislature for 300 pounds for three years to help defray expenses for such an endeavor. In 1801, despairing of either academic or government support, he decided to devote his own professional income to the project.

He purchased twenty acres, three and a half miles from the center of New York City, well adapted for "plants that require a peculiar aspect or situation." The land sloped to the east and south and "contained remarkable difference of soils, from the rocky upland to the hilly slope and the moist and watery bottom adapted to the cultivation of a great variety of vegetable specimens including grasses." He proposed that individuals begin to collect, in their own areas, "for the purpose of forming a complete Flora of the State."[4]

In 1810 the botanic garden (now called Elgin, after the Scottish place of origin of Hosack's father) was enclosed by a stone wall. The layout was artistic and ingenious and included a kitchen garden nursery for introducing choice fruits. "A belt of forest trees surrounds the establishment: the oak, the plane, the elm, the sugar maple, the locust, the horse chestnut, the mountain ash, the basket willow and various species of poplar are distributed and beautifully combine." Pines, junipers, yews, and hemlocks were planted on the rocky slopes. A big oak stood on the lawn. Different species of magnolia mingled with the fringe tree (*Chionanthus virginicus*), the thorny aralia (*Aralia spinosa*, whose bark possessed medicinal properties), and the snowdrop tree (*Halesia carolina*).

The greenhouse, built in 1803, was like William Hamilton's at Woodlands, near Philadelphia, sixty-two feet long, twenty-three feet deep, and twenty feet high. To Dr. Thomas Parke, Hosack wrote, "I shall heat it by flues. They will run under the stays so they will not be seen." He continued, "My walks will be spacious." By the following summer he hoped to have hothouse accommodations for plants from his friends "in Europe and in the East and West Indies," to whom he had written. He wished to know what medicinal plants Bartram could supply and hoped Hamilton would send him duplicates of his rare and valuable plants in exchange for "anything I possess." He embarked upon a system of trades with the Danish professor Martin Vahl and with Thouin of the Paris Jardins des Plants, from whom, through Jefferson, Hosack also received (like M'Mahon) seeds from the Lewis and Clark expedition.

But Hosack's ambitious project was short-lived.[5] By 1811 he could get no one to take over the garden from him, and it fell into disuse. His original vision was not realized until 1889, when the Torrey Botanical Club drew up plans for what is now the New York Botanical Garden. A plaque at the site of the original Elgin Botanic Garden honors David Hosack as "botanist, physician, man of science and citizen of the world." That early garden (1801–11) "for the advancement of medical research and knowledge of plants" was the progenitor of many similar endeavors across the country as the nation moved westward. The site lies directly beneath the present Rockefeller Center.

By 1828 Hosack had bought an estate of 700 acres on the Hudson, named Hyde Park in honor of an early British governor and recently owned by his old friend Samuel Bard, from whose son Hosack purchased it. The advertisement for its sale in 1768 read:

> filled with good timber . . . a great part of the upland exceeding good for grain or grass . . . a large well improved farm with a good house . . . a large new barn, a young orchard of between five and six hundred apple trees, most grafted fruit . . . between thirty and forty acres of rich meadowground fit for the scythe, and about one hundred and fifty acres of upland cleared and in tilling order . . . three good landing places . . . where the largest Albany sloop can lay close to a large flat rock.

Bard and his wife had planted extensively since their retirement there in 1798, and Hosack knew the place well from having assisted the Bards in procuring plants, from one Mr. Prince of Flushing and elsewhere.

To Hyde Park, then, Hosack repaired with his third wife and proceeded to make the estate into one of the showplaces along the Hudson. He had welcomed André Parmentier on his arrival in New York and invited his assistance in landscaping the country estate. Consequently, Hosack was the first to show the "natural" landscaping style in the New World.

For a moment, we will skip ahead through our chain, watching its continuity. Hosack is the first link; Parmentier the second. Another, Andrew Jackson Downing, left us descriptions of the property, for which he had only praise.

Downing describes the approach from the landing, where guests were met by Dr. Hosack's carriage. The drive wound up the incline toward the house and had been designed by Parmentier to move through woods to views of open fields with clumps of trees. Parmentier had also designed the walks, which were twenty-five to thirty feet wide and bordered with beds of mixed shrubs and herbaceous plants. These were planted with the smallest plants nearest the walk, the taller ones just behind them; back of the flower borders came plantings of low shrubs, backed in their turn by higher ones. The whole was set against a background of trees of moderate size. "The effect of this belt ȯn so large a scale, in high keeping," Downing wrote, "is remarkably striking and elegant." We shall see this idea for large-scale border plantings turning up later in descriptions of other gardens.

Another feature of lasting impact was described by the last link in the chain, Charles Eliot, who visited Hyde Park toward the end of the century. He was taken to see the planting of the trees along the roads—they stood formally, in double rows, where the roads ran straight; but where the roads curved, the trees were grouped informally, to afford shade and yet allow views across carpets of lawn.

ANDRÉ PARMENTIER, first of many who sought to introduce new methods of garden design to American householders, was a native of Belgium, born into a prominent family much involved in horticultural pursuits. One brother, Joseph, was superintendent of the great Parc d'Enghie, where André and his other brother, Louis, worked among the thousands of specimen trees and shrubs in the formal layout said to have been studied by Le Notre. A cousin was the Parmentier whose fame as a producer and popularizer of the potato remains with us today. Louis went on to become a rose fancier of note, with hundreds of varieties to his credit. André, burdened by financial losses at home, brought his wife and child and a large collection of plants across the Atlantic in 1824, with an idea of starting a career as a nurseryman in the West Indies.

On landing in New York he was persuaded to end his journey there, possibly by Dr. Hosack, who offered him the superintendency of the Elgin Botanic Garden. Parmentier preferred to set up for himself on a piece of wild land in Brooklyn. He cleared this and planted it with such industry that within three years he had a nursery business in shrubs and trees, an orchard of fruit trees, an exhibition garden, and an extensive vineyard. Within a few more years, he produced a catalogue and arranged for a team of agents to travel from Montreal to New Orleans, offering to sell from his lists a wide selection of the very best fruits—pears from Van Mons in the Netherlands (described as melting, buttery, baking), gooseberries from England, roses (presumably from his brother), strawberries (a new everbearing variety without runners), and innumerable other items. Parmentier is credited with introducing the "purple leaved beech" to American gardens and parks.

Parmentier laid out part of his land in a "picturesque" style, as described by a visitor from Baltimore in 1829, whose report will be quoted extensively in a moment. The visitor was particularly taken with the "rustic" seats dotted about, made from the bare and crooked limbs of trees, and with a "rustic arbour," from which one could look out over the gardens and the surrounding countryside. The visitor found these "to produce an agreeable effect by their natural appearance" and to be "in perfect harmony with the scenery around them." We shall see them turning up later across the countryside (and, still later, immortalized in cast iron).

In 1822 Thomas Green Fessenden began publishing *The New England Farmer,* "containing Essays, Original and Selected, relating to Agriculture and Domestic Economy with Engravings and the Prices of Country Produce."[6] For us, proof of his editorial acumen lies in his request for an original article from André Parmentier. He subsequently published the first exposition of the natural and picturesque styles of gardening to be made in the United States. Entitled "Landscapes and Picturesque Gardens," it is an important document that requires quotation in full (it can be found at the end of this chapter).

Our description of Parmentier's garden, with a plan, comes from Loudon's *Gardener's Magazine,* in volume 8 of 1832, and was written by the American visitor mentioned earlier, "J.W.S.," who toured the establishment in September 1829.

Parmentier's Garden, near Brooklyn. Sir, at the request of some of your readers in this country, I have compiled from different authorities, but chiefly from the *American Farmer,* an account of one of the first botanic gardens which has ever been established in this country, viz. that of Parmentier, about two miles from Brooklyn, Long Island. The following map . . . will serve to convey some idea of the general disposition of the whole; but I am confident that neither plan nor description can furnish any adequate idea of the particular beauties of the place. . . . [T]he garden of M.

29

Links 100

0 1 2 3 4 5 *ch.*

a, Dwelling-house.
b, Labourers' dwellings, 2.
c, Tool and work-house, 2.
d, Barn.
e, Green-houses, 2.
f, Hotbeds, 3.   g, Plan
    for plants in summer, 2.
h, Herbaceous plant gar-
    den.   i, Rustic arbour.
k, French saloon.
l, Nectarine and peach
    tree alley.
m, Pear tree alley.
n, Apple tree alley.
o, Plum tree alley.
p, Cherry tree alley.

1. Vines, 10 squares, 263 kinds.
2. Rose trees, 2 squares, 250 kinds.
3. Ornamental trees, 7 squares.
4. Peach trees, 4 squares, 64 kinds.
5. Apple trees, 3 squares, 24 kinds.
6. Plum trees, 2 squares, 85 kinds.
7. Pear trees, 4 squares, 190 kinds.
8. Cherry trees, 2 squares, 71 kinds.
9. Imported fruit trees, 5 squares.
10. Young vines, 5 squares.
11. Quince stocks, 1 square.
12. Monthly strawberries, 1 square.
13. Place for manure and weeds.
14. Jamaica turnpike.
15. Flatbush turnpike.

The other kinds of fruit are : — Nectarines, 15 kinds; apricots, 18 kinds; walnuts, chestnuts,
filberts, and hazel nuts, each 20 kinds; quinces, 5 kinds; raspberries, 5 kinds; gooseberries,
20 kinds; currants, 7 kinds; strawberries, 17 kinds.

Parmentier's garden, plan and key. From "Foreign Notices: North America,"
in *The Gardener's Magazine,* edited by J. C. Loudon, 8 (1832): 71. Courtesy
Melanie Simo.

Parmentier is, perhaps, the most striking instance we have of all the different departments of gardening being combined extensively and with scientific skill. The rapidity with which this garden was formed added to its effect. Nearly twenty-five acres of ground were originally enclosed; and the inhabitants of the vicinity beheld with astonishment, in the short space of three years, one of the most stoney, rugged, sterile pieces of ground on the whole island, which seemed to bid defiance to the labours of man, stored with the most luxuriant fruit, and blooming with the most beautiful flowers.

The ground-plan of the garden, although without any remarkable inequalities, has yet some diversity of surface. The most elevated part, facing the south and southwest, is appropriated for the purpose of a vineyard; and several valuable varieties of the grape, foreign as well as indigenous, are there cultivated. The beds of the ornamental part compose broad belts laid out in a serpentine direction, and edged with thrift (Statice Armeria). These sections contain a mixture of plants and shrubs of both the Old and the New World. The several species of Robinia, the Philadelphus grandiflorus, the Halesia, the Pteleae, and many others conspicuous for their beauty, are interspersed and contrasted with the delicate Tamarix of Europe; the paper mulberry, now bearing its curious fruit; several species of shrubby willows and poplars; the splendid Anchusa capensis, with its azure blossoms; the no less luxuriant Balsamina; and thousands of others which we might mention, all disposed in the most artful manner, so as to heighten the effect, and yet to conceal too glaring an appearance of art.

In the northern parts of the garden are nurseries, containing young plants of every kind of tree which is to be found in the beds. To the left of the garden, an avenue leads to a rustic arbour, in the grotesque style, constructed of the crooked limbs of trees in their rough state, covered with bark and moss: from the top of this arbour a view of the whole garden and the surrounding scenery is obtained; including Staten Island, the Bay, the Governor's Island, and the city of New York. At some distance from the rustic arbour is a plot of ground, called the French Saloon; a beautiful oval, skirted with privit (Ligustrum), kept dwarf to the height of

1 ft., and enclosing a solid mass of China monthly roses. The various kinds of fruit trees are carefully arranged, and the alleys leading to them are skirted with specimens of the different sorts in a bearing state, for better exhibition, and to furnish the necessary grafts for the establishment.

The green-house department, although not so extensive as some other parts of the garden, contains many beautiful plants, exhibited with the same tasteful arrangement which characterizes every part of M. Parmentier's establishment; and which displays itself even in the grouping of the pots, which are all arranged according to the colour and size of the flowers: thus showing the variety of ways in which a skillful gardener may distribute his materials to produce a picturesque effect.

The manner of protecting the plants in this garden from the violence of the weather or the heat of the sun is quite novel in this part of America; canvass covers being so managed as to be rolled or unrolled with the greatest ease and despatch, by means of ropes and pulleys. The necessity of some such screen is quite obvious, when plants, and particularly tender exotics, are exposed to our excessive sun, and yet it is too generally neglected among our gardeners.

In short, this establishment is well worthy of notice as one of the few examples in the neighbourhood of New York of the art of laying out a garden so as to combine the principles of landscape-gardening with the conveniences of the nursery or orchard.

In *A Treatise on the Theory and Practice of Landscape Gardening* (1844), Andrew Jackson Downing reported on Parmentier's importance to the advancement of landscape gardening in America.

During M. Parmentier's residence on Long Island, he was almost constantly applied to for plans for laying out the ground for country seats by persons in various parts of the Union as well as in the immediate proximity of New York. In many cases he not only surveyed the demesne to be improved but furnished the plants and trees necessary to carry out his designs. Several plans were prepared by him for residences of note in the southern states, and two or three places in Upper Canada, especially near Montreal, were, we believe laid out by his own hands and stocked from his nursery grounds. In his periodical catalogue he arranged the hardy trees and shrubs that flourish in this latitude in classes, according to their height, etc., and published a short treatise on the superior claims of the natural over the formal or geometric style of laying out grounds. In short, we consider M. Parmentier's labours and example as having effected directly far more for landscape gardening in America than those of any other individual whatever.

Note: As M. Parmentier's article is a milestone, and as the function of a milestone is to be right on the line, we include it here in its complete text.

Landscapes and Picturesque Gardens
by Mr. A. Parmentier, of New York

It has been reserved for the good taste of our age to make many advantageous changes in the embellishment of gardens, and to re-instate Nature in the possession of those rights from which she has too long been banished by an undue regard to symmetry.

Our ancestors gave to every part of a garden all the exactness of geometric forms: they seem to have known of no other way to plant trees, except in straight lines; a system totally ruinous to the beauty of the prospect. We now see how ridiculous it was, except in the public gardens of the city, to apply the rules of architecture to the embellishment of gardens.

The majestic trunk is now allowed the liberty of displaying its form, or of following in its vigorous shoots the plan of nature. Gardens are now treated like landscapes, the charms of which are not to be improved by any rules of art.

The advantages of these changes are so manifest, and so well appreciated, that further proofs seem unnecessary. For where can we find an individual, sensible to the beauties and charms of nature, who would prefer a symmetric garden to one in modern taste; who would not prefer to walk in a plantation irregular and picturesque, rather than in those straight and monotonous alleys, bordered with mournful box, the resort of noxious insects?

Where is the person, gifted with any taste, who would not choose these alleys that wind without constraint, in preference to those dull straight lines which can be measured by one glance of the eye, and the monotony of which is unvaried? Instead of this, the modern style presents to you a constant change of scene, perfectly in accordance with the desires of a man who loves, as he continues his walk, to have new objects laid open to his view. To understand this style of a garden requires a quick perception of the beauties of a landscape, without which the existing plantation might be destroyed, instead of being used. Limited prospects, and neighbouring houses and buildings not worthy of notice, should be concealed, and the view left open to those objects which strike the eye of the beholder agreeably. The front of the house ought always to be *uncovered,* the principal point of view seen or conjectured. A vast idea of the proprietor should be given, and this can only be done by a grand plan, in which nothing niggardly is to be seen.

Rows of trees should never be planted in front of the house, particularly when the house has been built in good taste, and at great expense. It may be objected to this, that the shade is wanted, and this I would not exclude; but, instead of one row in front, I would plant thick groups of trees on the three other sides, and leave the front open to public view; otherwise the taste and expense are, in great measure, thrown away.

When you choose the situation of a country-house, let it be at some distance from the public road, so that the road which leads to it may give a good idea of the extent of the proprietor's domains, and care should be taken that the road is proportioned to this extent. It ought to be from eight to ten feet wide, so that carriages may pass, and gently serpentine. This winding should have a reason—that is to say—some groups of trees should be so placed as to appear to be the cause of it: for naturally the road would have led directly to the house, but the person walking, when he observes these groups of trees, will see at once why it does not. Besides, he will be agreeably amused by the variety of views which will show to advantage the manner in which the artist has executed his plan, and the choice he has made of a situation.

If the house is placed on an eminence or side-hill, the prospect will be much more beautiful if you can enjoy the view of water:

and, to add to the whole effect, and facilitate the labours of the artist, it is desirable that a grass plot should naturally present itself.

The plantations and groups of trees near the house should be, if possible, of a deeper green;—they would extend the view and the perspective, and produce the effect of shades in a landscape-picture, where the groups of trees in front are of a darker shade, and seem to remove the perspective from the extremity of the landscape. For the same reason, the trees at the farther part of a park, or garden, should be those of a thin and light foliage.

Plantations should consist of, besides merely ornamental trees, those fruit-trees which are high and of bright foliage. Their flowers in spring, and branches loaded with fruit in autumn, make them objects of great beauty and interest.

The apple-tree alone, on account of its horizontal branches, should be confined to the orchard, where its useful products are ornamental and valuable. The most should be made of the agreeable and interesting views which may be had in the neighbouring landscape. They may be made useful to the general plan by being represented as the property of the proprietor.

For this reason, I highly approve of blind fences, and live hedges. But fences, necessary as enclosures, should be concealed so as not to appear as boundaries to the establishment, and present to the eye a disagreeable interruption in the prospect. The judicious use of hermitages, arbours, cottages and rotundas will add to the effect, in picturesque gardens and ornamented farms. If you use these ornaments, place the hermitage in some retired spot: a small rivulet would be appropriate and beautiful accompaniment. The rustic arbour and cottage may occupy a place less secluded. An elegant rotunda should be seen from a distance, and on a hill or eminence. It should make a part of the establishment of a wealthy man, as well as pagodas, turrets, and Chinese towers. These ornaments are so expensive, that they are beyond the means of most persons: but hermitages, arbours, and cottages may generally be afforded, as there is little expense in their construction.

Rustic bridges are very pretty where there is a stream, and they can be made of use; but they have no pretensions as mere accompaniments to a plantation. Obelisks, columns, &c, should be placed on elevated places.

As to tombs and cemeteries, I should wish to banish them entirely from gardens. They always awaken melancholy reflections in old people, for they remind them of their approaching end; and a regard for their feelings should, I think, exclude from their places of resort every object which could have such an effect.

Whilst on this subject, I will mention an anecdote of the celebrated Kent, architect of the English gardens, which will show to what extent this mania may be carried. He built a tomb in a park, and, to make the place still more gloomy, planted around it dead and mutilated trees; but, notwithstanding the celebrity he had acquired, he was loaded with ridicule, and forced to displace the trees.

NOTES TO CHAPTER 8

1   Dr. Bigelow accompanies the earlier individuals; Scott follows Downing in home gardens, if not in parks; nurserymen and seedsmen of the century will be present with their offerings; and George Perkins Marsh sounds, in midcentury, the first call for conservation. But the major developments follow the six individuals of our framework.

2   With all his concern with the events of his time (he was even involved professionally in the Burr–Hamilton duel), Hosack managed a busy social life, married three times, and raised a large family.

3   One of his first pupils was DeWitt Clinton, who remained a lifelong friend and also became one of the American notables to be called upon by foreigners visiting the new nation.

4   In 1792 a New York Society for the Promotion of Agriculture, Arts, and Manufactures had been founded, to which Hosack had donated his collection, made in England, of sixty species of grasses, "carefully prepared and arranged with botanic and common English names." This society also funded a professorship in natural history, occupied by Dr. Samuel Mitchill (in addition to his position at Columbia College).

5   The most rewarding fruits of Hosack's botanic innovations can rest with two men: John Torrey and Asa Gray. Torrey had assisted Hosack in his botany courses in 1822 and went on to write the *Flora of North America* in collaboration with Gray, who, at seventeen, had bought the *Manual of Botany* written by Amos Eaton and had begun collecting. After attending medical school, Gray arrived in New York with a letter of introduction to Torrey, who took him on as an assistant. (Hosack also acquired other distinguished witnesses to his efforts. He had taken in the wandering botanist Frederick Pursh and even considered starting a botanical magazine, based on Curtis's, with Pursh as editor.)

An involved but interesting story is pertinent here. Eaton had been a student of Mitchill and Hosack and had been sentenced to jail because of errors he made as a land agent. William Torrey, the state prison's fiscal agent, employed him, and Eaton became a friend of Torrey's son, John, through their mutual interest in botany. Eaton wrote to his wife, "I have contrived a new method of arrangement by which I can exhibit all the known species of plants (about forty thousand) in one small duodecimo volume. So that you can readily determine the name of every plant and its uses in medicine, diet, agriculture and the arts by merely inspecting the flower and some few other parts. The Agent's son, John Torrey, supplies me with the books I want which are not in the state prison library." After serving his sentence, Eaton went out into the world again, to lecture at Yale and Williams with such success that Asa Gray, soon to be recognized as the leading American botanist, credited him with popularizing the teaching of natural history. Eaton's *Manual of Botany*, published in 1817, ran into eight editions.

6   As customary in the nineteenth century, the presentation of new ideas in American life was accompanied by protestations of extreme modesty, however false. Pleas of friends, difficulties of distance, discovery of a void, a wish to help one's fellow men—all had to be stated. These are implicit in the introduction to *The New England Farmer,* where Fessenden protests that two previously published and popular magazines (*The Ploughboy,* published in Albany, and *The American Farmer,* published in Baltimore) "cannot be so conveniently circulated in New England." So—for two dollars and a half in advance, or three dollars at the end of the year—one was invited to subscribe to his new effort.

NINE

Cemeteries into Parks

HE conjoining of gardens, parks, and cemeteries may seem far-fetched in a study of the making of nineteenth-century American gardens, but it resembles what Cotton Mather called the practice by one individual of both religion and medicine, "a divine conjunction."

As early as the beginning of the nineteenth century, it had become evident, particularly in old cities, that the conventional plan of burying the dead in congested areas would no longer suffice. In England, where burial places were hallowed ground in churchyards, the problem had become acute. New burials took place on top of older ones, and eventually the space became too crowded for either public health or private solace. Captain Hall described the typical English burial ground as "a soppy churchyard, where the mourners sink ankle-deep in a rank and offensive mould, mixed up with broken bones and fragments of coffins," and doubted that any virtue could be derived from "the recollections of coughs, colds and rheumatism out of number, caught whilst half a dozen old fellows with longtailed threadbare black coats are filling up a grave for which they themselves might seem the readiest tenants."

Americans seemed to have had a special fondness for cemeteries as places to visit, to court in, and to show foreign guests. On an October afternoon in 1827 Captain Hall was taken to visit the cemetery in New Haven. His impressions were favorable.

[W]e drove out of town to the Grave-yard, one of the prettiest burying places I ever saw. It occupies . . . twenty acres laid out in avenues and divided by rows of trees into lots for the different inhabitants. These connecting lanes or roads are not gravelled but laid down in grass as well as the intermediate spaces which are spotted over with handsome monuments of all sizes and forms, giving a lively instead of a gloomy air to the whole scene.

There is certainly some improvement in this compared with the practice of huddling together so many graves in the confined space round the places of worship in a populous city.

In the New England towns, the best spot with the loveliest view was set aside for a burying ground. But burial of the dead remained a personal matter, as witness the small family cemeteries on the outskirts of gardens and fields down the rural East Coast, from Maine farms to Virginia plantations. As Mrs. Trollope noted, "*In Virginia and Maryland* almost every family mansion has its little grave-yard, sheltered by locust and cypress trees," although she also added, "and one mansion on the Delaware, near Philadelphia, has the monument which marks the family resting-place, rearing itself in all the gloomy grandeur of black and white marble, exactly opposite the door of its entrance."

Partly because of the macabre journalistic onslaught on old London cemeteries, and partly as a result of commonsense village-into-city forecasts in America, there was a move toward improving American cemeteries to the point of making them attractive even to the living—in fact, of treating them as parks. As with many of the social reforms for which Americans became famous in the nineteenth century (such as democratic government, universal public education, prison reform, and public health measures), the initial impulse came from Boston.

IN Boston the first graveyards had been established to receive and commemorate the distinguished fathers of the new colony, the governors and magistrates in family lots surmounted by long rectangular stone monuments, like coffins or stone tables on which family names could be inscribed, or by tombs like small houses dug into the hillside.

Simple, impressive, but doomed to be overcrowded. In Boston also, the medieval custom of burial under churches had been followed where church affiliation had been established by the Church of England. Heroes of both sides at Bunker Hill lie today in the Old North Church.

In 1825, Dr. Jacob Bigelow of Boston, already famous for his contributions to medicine, education, and horticulture, and our earlier acquaintance, was approached by a group of concerned philanthropic citizens about a new idea for a parklike cemetery.[1] Bigelow's own history of the ensuing project tells us that his "attention had been drawn to some gross abuses in the rites of sepulture as they then existed under churches and in other receptacles of the dead." At this time, he says, his own "love of country, cherished by the character of earlier pursuits," had led him to

> desire the institution of a suburban cemetery in which the beauties of nature should, as far as possible, relieve from their repulsive features the tenements of the deceased; and in which, at the same time, some consolation to survivors might be sought in gratifying, as far as possible, the last social and kindred instincts of our nature.

Within a year, Dr. Bigelow called a meeting of "a few gentlemen" at his house, where he submitted to them a plan for a cemetery composed of family burial lots, separated by and interspersed with trees, shrubs, and flowers, in a wood or landscape garden. This was approved and a committee appointed to "look out for a tract of ground suitable for the desired purpose." Several proposed sites were unattainable, because of their high price or from "the reluctance of the owners to acquiesce in the use proposed." After nearly three years, a tract in Cambridge and Watertown known as Stone's Woods, though "more familiarly to Harvard College as 'Sweet Auburn,'" became available. It was owned by Mr. George W. Brimmer, "a gentleman whose just appreciation of the beautiful in nature had prompted him to preserve from destruction the trees and other natural features of that attractive spot."[2]

In 1830 Dr. Bigelow proposed to enlist sufficient subscribers to purchase the whole area for an "ornamental cemetery." Mr. Brimmer generously offered to accept only what the land had cost him and became one of the "most active members of the first Committee or Board of Managers."

Divine conjunctions being what they were in Boston in the early days, Dr. Bigelow had been chosen corresponding secretary of the newly incorporated Massachusetts Horticultural Society the year before the land was acquired. "At that time," wrote Dr. Bigelow, "there was no ornamental rural cemetery, deserving of notice, in the United States, and none even in Europe, of a plan and magnitude corresponding to those which Mount Auburn" was to possess. However,

> the subject was new, the public were lukewarm, and in many cases the prejudices and apprehensions of the community were strongly opposed to the removal of the dead from the immediate precincts of populous cities and villages to the solitude of a distant wood.

Dr. Bigelow realized the value of a "young, active and popular society" in overcoming these prejudices. The Horticultural Society, young and destitute, found itself officially raising the money to pay Mr. Brimmer. "To accommodate the wishes of the horticulturists, an experimental garden for the cultivation of flowers, fruits, etc." was allowed for in the allotment of the ground. After a year of exploratory visits to Sweet Auburn and several explanatory leaflets, a scheme was proposed, "in behalf of the Horticultural Society," to purchase land from the patient Mr. Brimmer "as soon as one hundred subscribers for cemetery lots at sixty dollars each should be obtained." By the spring of 1831, the Horticultural Society closed the deal for the "experimental Garden and Cemetery" (note the order) and asked that the establishment, including the garden and cemetery, be supplied a definite name. Mount Auburn it became.

As Mount Auburn was to be a rural beautification achievement, incorporating the natural advantages of the site to make a place of recreation for the living as well as of repose for the dead and of comfort for mourners, it would take a man of many parts to accomplish the project. Dr. Bigelow was easily the best-qualified man of the day in Boston— a practicing physician, a botanist, popular with his fellows, and an artist. A look at the original plan will show the originality of his inspiration.

Landscaping for the project rejects the usual grid and develops a loosely woven net of avenues and paths, draped over the rising ground and incorporating ponds in the lowland area.[3] Virgil, in his *Georgics,* recommended following the contours of the land for plowing (very likely many of the lot purchasers could quote this passage by heart). From the enormously impressive Egyptian entrance gate, said to have been designed by Dr. Bigelow (who seems to have been credited with the design of everything, horticultural or architectural, in the cemetery), there is an open space on the right of the entrance for a lawn, which lies before the granite Gothic chapel, also popularly attributed to Dr. Bigelow. (As we are not writing a guidebook to the cemetery, beyond attesting to the idea of Dr. Bigelow as a genius, we must leave correcting attributions to others.) To the left of the entrance stretches a large expanse, uncut by paths, which, because it is bordered by Garden Avenue, must have been set aside for the experimental garden, though this never transpired "for want of specific funds" for its support "and various other causes." At the top of the highest hill is a tower (another of Dr. Bigelow's supposed designs, not unlike a simplified Qutb Minar in Delhi), which was named for Washington. A sphinx, originally commissioned by Dr. Bigelow, awaited an appointed place and function until it finally became the monument to honor the Civil War dead in 1872. Four large statues were commissioned to stand near the Gothic chapel. The first was of Justice Joseph Story of the United States Supreme Court and was paid for by his friends. Three others— of John Winthrop, James Otis, and President John Adams—were paid for by the board.

Mount Auburn became such a showplace that by 1839 a guide for visitors was issued, called *The Picturesque Pocket Companion.* Perhaps the experimental garden never evolved because the whole area became a garden. It was claimed the soil was not suitable for an experimental garden. In any case, the Horticultural Society in no way felt betrayed and years later realized a handsome sum—about forty times the sum paid Mr. Brimmer—in settlement of its withdrawal from the agreement.

BY 1849 the relationship between cemeteries and parks was pointed up by Andrew Jackson Downing in *The Horticulturist.* He hailed as "one of the most remarkable illustrations of the popular taste in this country . . . the rise and progress of our public cemeteries." Twenty years

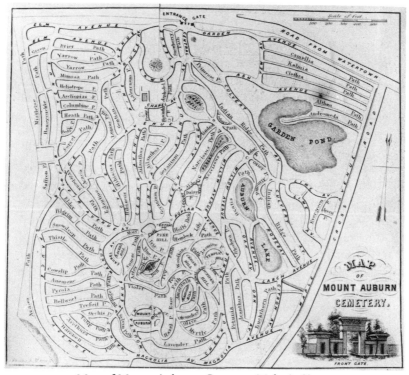

Map of Mount Auburn. Courtesy Melanie Simo.

earlier, he said, they had been nothing better than common graveyards, except for a few—like the burial ground at New Haven, where
Downing, unlike Captain Hall, found little to praise, "only a few
willow trees" to break "the monotony of the scene." However, when
Mount Auburn was made a rural cemetery, eighteen years before
Downing wrote his assessment, the idea of "the charming natural site,
finely varied in surface, . . . admirably clothed by groups and masses
of forest trees . . . tastefully laid out . . . monuments . . . the whole
highly embellished," took the public mind by storm. "Travelers made
pilgrimages to the Athens of New England solely to see the realization
of their long cherished dream of a resting place for the dead, at once
sacred from profanation, dear to the memory, and captivating to the
imagination."

In 1849 Downing asserted that scarcely a city of note in the whole country did not have its rural cemetery.

> The three leading cities of the north, New York, Philadelphia, Boston, have, each of them, besides their great cemeteries— Greenwood, Laurel Hill, Mount Auburn—many others of less note . . . any of which would have astonished and delighted their inhabitants twenty years ago.

Here he makes a point that was to change American cities for all time.

> The great attraction of these cemeteries is not in the fact that they are burial places . . . all these might be realized in a burial ground planted with straight lines of willows and sombre avenues of evergreens. The true secret of the attraction lies in the natural beauty of the sites, and in the tasteful and harmonious embellishment of these sites by art. . . . Hence to an inhabitant of the town a visit to one of these spots has the united charm of nature and art—the double wealth of rural and moral associations. . . . Indeed, in the absence of great public gardens, such as we must surely one day have in America, our rural cemeteries are doing a great deal to enlarge and educate the popular taste in rural embellishment.

Interestingly, Downing continued, although these three leading examples are all laid out in admirable taste, with the greatest variety of trees and shrubs to be found in the country, and kept in a manner seldom equaled in private places, they differ in their essential characters. Greenwood in Brooklyn is large, grand, dignified, and parklike, laid out in a broad and simple style and commanding ocean views. Mount Auburn is richly picturesque in varied hill and dale and owes its charm mainly to a variety of sylvan features. Laurel Hill near Philadelphia is a charming pleasure ground, filled with beautiful and rare shrubs and flowers, "at this season a wilderness of roses, as well as fine trees and monuments." In a footnote, Downing protested the "hideous iron-mongery" beginning to disfigure cemeteries with elaborate fences, gates, and, most deplorably, coats of arms.

But this last was only a footnote; after describing the roses and monuments he began to try to persuade his readers. He directed them to understand the influence these beautiful cemeteries constantly exercise on the public by considering how rapidly parklike burial settings had increased in fifteen years. The numbers of visitors they attracted was an indication of the extent to which they had aroused public interest. Laurel Hill, four miles from Philadelphia, had counted 30,000 visitors in one year; double that must have visited Greenwood; Mount Auburn must certainly have had an equal number. Rural cemeteries, in the absence of public gardens, were filling their place to a certain degree. Downing suggested that public gardens established in a liberal and suitable manner near large cities would be equally successful, that they would rapidly educate the public taste, and that the progress of horticulture as a science and as an art would be equally benefited.

Downing added,

> The passion for rural pleasures is destined to be the predominant passion of all the more thoughtful and educated portion of our people, and any means of gratifying their love for ornamental or useful gardening will be eagerly seized by hundreds of thousands of our countrymen.

We are here watching the birth of an American park, Central Park to be its name.

> Let us suppose a joint stock company, formed in any one of our cities, for the purpose of providing its inhabitants with the luxury of a public garden. A site should be selected . . . [to] have a varied surface, a good position, sufficient natural wood, with open space and good soil enough for the arrangement of all those portions which required to be newly planted.

Downing envisaged its bright future. In 50 to 100 acres, an example can be afforded of laying out grounds, thus teaching practical landscape gardening. A collection of all the hardy trees and shrubs that grow in this climate, each distinctly labeled, will insure that the most ignorant visitor can learn something of trees. A botanical arrangement of plants and a lecture room would allow for educational activities. A magnificent wooded drive could be laid out, and (in the same sentence) suitable ices and other refreshments could be served, as in the German gardens, which, along with the finest music and the most rigid police, would tempt the better classes to such a resort. In fact, it would be "the greatest promenade of all strangers and citizens, visitors or inhabitants of the city of whose suburbs it would form a part."

The park could be supported by a small admission fee and by subscription. Only shareholders, like those who own lots in a cemetery, would be allowed to bring in their horses and carriages—a privilege that would tempt hundreds to subscribe. No traveler could leave the city without seeing such a public garden, the city's most interesting feature.

Two points remained in Downing's argument. If the road to Mount Auburn was lined with coaches carrying thousands and tens of thousands, then a garden full of varied instruction, amusement, and recreation should be ten times more visited. And if hundreds of thousands in New York pay to see stuffed boa constrictors at Barnum's Museum on Broadway and will incur the expense of going six miles to see Greenwood, surely one could safely estimate that many more would resort to a public garden.

Besides being profitable, the garden would civilize and refine the national character, foster the love of rural beauty, and increase the knowledge of and taste for rare and beautiful trees and plants. If only one of the three cities that first opened cemeteries would set an example, the practice of making such public spaces would become widespread. The true policy of republics, Downing argued, is to foster the taste for great public libraries, sculpture and picture galleries, parks and gardens that *all* may enjoy.

So, thanks to Mr. Downing, within a very few years Central Park was established in New York City. Before considering its development, we would do well to understand more about the history and theory of landscape gardening in the nineteenth century, which we can do by getting to know the works of John Claudius Loudon, the English authority, and Andrew Jackson Downing, one of the century's most influential American personages—and the next link in our chain.

NOTES TO CHAPTER 9

1 Dr. Bigelow was a man of many concerns and skills. In 1814 he had published a book called *Florula Bostoniensis; or, A Collection of the Plants of Boston and Its Vicinity,* intended "for the use of a botanical class in the city," because there was a "great deficiency

of books relating to American plants." At the same time, collaborating with Dr. Francis Boott, he began a collection of materials for a *Flora* of the New England states and made several botanical tours, into the mountains of New Hampshire and Vermont and along the seacoast, for material for this enterprise, which they later relinquished as it was "rendered more difficult by other engagements." One of these must have been the 1817 publication of Bigelow's three volumes of an *American Medical Botany, being a Collection of the Native Medicinal Plants of the United States,* with colored engravings of sixty plants from original drawings made by the doctor himself. By the time the second edition of the *Florula* appeared in 1824, greatly enlarged, the need for a New England botanical work was less pressing, according to Dr. Bigelow, works having appeared on American botany by "Muhlenberg, Pursh, Elliott, Nuttall, Eaton and Torrey." Nevertheless, Dr. Bigelow had incorporated into the second edition of the *Florula* all the material he had collected for the proposed New England work, doubling the size of the book. Using the Linnaean system of identification, he disarmingly sympathized in the introduction with the "students of science" who had to "unlearn continually what they had acquired," due to the "continually shifting nomenclature of plants." Although the second edition left out the actual locations of examples of plants given in the first edition, the book remained engagingly provincial, with notes aiding identification, comparing engravings of English specimens, and supplying helpful hints as to uses: for instance, *Salicornia,* "different species of which are among the maritime plants employed in the manufacture of soda. They are used at table as pickles."

2   Later Mr. Brimmer increased his original purchase to include land along the highway where the cemetery entrance now stands.

3   In the loosely woven net the largest strands were "avenues" named for trees—Elm, Willow, Rosebay, Cedar, Cypress, Spruce, Magnolia, Larch, Beech, Lime, Oak, Chestnut, Poplar—except for two, practically named Central and Culvert. Finer strands were "paths" named for all the most familiar small shrubs and flowers—Hibiscus, Ailanthus, Lupine, Gentian, Heath, Eglantine, Yarrow, Mimosa, Heliotrope, Asclipias (*sic*), Snowdrop, Violet, Anemone, Bellwort, Pyrola, Trefoil, Tulip, Arethusa, Harebell, Orchis, Geranium, Narcissus, Lily, Daisy, Myrtle, Lavender, Acacia, Verbena, Aster, Sumac, Moss Holly, Kalmia, Clethra, Camellia, Andromeda, and Primrose. Thistle and Briar are honestly added, and, surprisingly, Hemlock. Pilgrim is next to Elder, making one wonder if Elder is human or horticultural, until one remembers the revered "virtues" of the shrub. The dells are named Hazel and Consecration.

TEN

Principals and Their Principles

HERE were two influential agents in the conveyance to the American scene of information on the enormous changes taking place in English and French landscape gardening and in the recording of the American contribution to the art. In our pursuit of the true character of Americans and their gardens in the nineteenth century, we shall do well to lay out and compare the principles set forth by the two leading landscape gardeners of the time, one English, one American. In England, John Claudius Loudon dominated the scene, where writers contended for the last word on gardening subjects, nervous or restless owners hired first one landscape gardener and then another, and opinionated, self-appointed authorities crowded into the gardens to nip at each others' heels. In the United States, the race was relatively open and unset until Andrew Jackson Downing became the first to hold forth on artistic and philosophical American approaches to the national occupation of home building and landscape beautification.

Though America remained, due to distance and its preoccupation with taming a wilderness, fairly immune to direct foreign influences in garden design and landscape planning, books from England and France made good reading in American libraries. For those Americans with an awareness of natural beauty, the detection of the "picturesque," discovered or created, afforded material for discussion, if not for actually changing the landscape. John Adams immortally recorded the American point of view on new European landscape designs after he toured the leading estates of England with Thomas Jefferson in 1786. Jefferson had been greatly impressed by the English skill in gardening, but Adams was full of patriotic reservations.

[I]t will be long, I hope, before Ridings, Parks, Pleasure Grounds, Gardens, and ornamented farms grow so much in fashion in America. But Nature has done greater things and furnished nobler materials there. The oceans, islands, rivers, mountains, valleys, are all laid out upon a larger scale.

And although Abigail Adams had learned to judge a foreign landscape by the contributions made to it by "nature's handmaiden, art," she also felt no need to emulate at home what she had seen abroad.

The spirit of landscape design was beginning, even then in England, to relax. Humphrey Repton suggested "improvements" of estates to introduce a lightness, an invitation to joy, even a little gentle humor. In his famous "Red Books," sketches of "before" and "after" could be flipped up and down to persuade the estate owner to introduce greater personal pleasures into his grounds, with the little figures of men and animals in each "after" sketch seeming to be in higher spirits than those in the view "before." In one case, treating a "sombre" view from a great house toward a wooded hillock, Repton presents the same hillock crowned by a "little cottage in the distance," which he feels "enlivens the view from the castle." The human element began to be included in landscape again as a matter of interest.

In America, Downing was enormously impressed by Repton's writings. We can see Repton's influece in the design for Downing's library. Repton, concerning himself with the design of houses as well as of landscaping, considered light to be all-important. A crosslight, he says, lends cheer, and windows opposite the fireplace spread light over the entire area—and the light, in turn, can be increased by a looking glass over the chimney.

We present here also a pair of pictures from Repton's *Fragments of the Theory and Practice of Landscape Gardening,* published in 1816, which illustrate the general enlivening of the formerly rather oppressively formal scene. In the pictures, an "ancient cedar parlour not much used" is shown in contrast to the "modern living-room," complete

The Cedar Parlor's Formal Gloom, and A Modern Living Room, both from
Humphry Repton, *Fragments of the Theory and Practice of Landscape Gardening*
(London, 1816). Both photographs courtesy Hunt Institute for Botanical
Documentation, Carnegie-Mellon University, Pittsburgh, Pennsylvania.

with an attached greenhouse. These two pictures and the accompanying poem mark a new approach to domestic living, the emergence of comfort and simple pleasures as desiderata.

Another reference collected from *Fragments* that appealed to Americans was to Longleat in Wiltshire, a landscape designed by Capability Brown, Repton's predecessor. As preferable to the old geometric formalities, Brown chose the gentler bucolic charms of acres of mown grass interrupted only by a distant artificial river, trees in carefully studied clumps, and deer grazing instead of less elegant cows. Repton nevertheless undertook the "improvement" of Longleat with enthusiasm.

> This magnificent park, so far from being locked up to exclude mankind from partaking of its scenery, is always open and parties are permitted to bring their refreshments; which circumstance tends to enliven the scene, to extend a more general knowledge of its beauties to strangers, and to mark the liberality of its noble proprietor.

George Washington had made a gardener's guidebook of Batty Langley's *New Principles of Gardening,* which had also guided his father-in-law, John Custis of Williamsburg. It was far grander in its concepts than anyone's visions for Mount Vernon, but Washington was pleased to adapt its language and some of its recommendations—as for "shrubberies"—to his own plans.

Jefferson had enthusiastically recommended Whately's *Observations on Modern Gardening* for every gentleman's library even before he bought a copy abroad to use while touring English gardens with John Adams, who had his own copy. Whately had absorbed the principles that treated landscape gardening as an art and had arranged them in order of practical application among great English country estates. Feature by feature (trees, rock, water, and so on), he conducted the reader to individual landscapes where these principles had been most grandly realized. Jefferson was also interested in costs and methods and tried to extract figures, as well as details of water-management systems, from the workmen he encountered. Jefferson approved the placement of temples and obelisks; John Adams, on the other hand, deplored most of the temples' patron gods and the absence of gentlemen from their own estates. He was so little impressed by the urge of the English landscape school to apply the tenets of art to nature that, as we have seen, he hoped it would never reach the shores of his native land, where, he maintained, nature had more to offer than art could possibly improve. The very composting practices of his own farm in Braintree were superior, he felt, to those he saw followed in England, even by the London city services.

By the end of the first quarter of the nineteenth century, the American goal of universal education was beginning to bear fruit, and the complaint of having "nothing to read" was real. With no copyright laws and less manners, American printers were able to get out copies of whatever they fancied as soon as the originals had been landed from the latest ship. A generation before, George Washington, writing to Arthur Young, the great English authority on farming methods, had observed that *our* farmers could read. Think, then, of the market for practical advice on farming and gardening, occupations that then engaged over half of the American population. When American seedsmen and nurserymen could burst into print on the eastern shores, run into dozens of editions in a few years, and move westward like a breeze, any horticultural magazines appearing in England were almost instantly absorbed into the American scene.

One measure of the infiltration of European ideas into American life is offered by a letter from Downing to Loudon published by Loudon in volume 15 of his *Gardener's Magazine* in 1839. In response to the publication of a series of encyclopedias compiled by Loudon, Downing summarized, with stark honesty, the state of American publications, horticultural exhibitions, villa gardening, apple growing, the still undiscouraged culture of the silkworm, and the emergence of a totally scientific and professional work on American botany by our two eminent botanists, Asa Gray and John Torrey.

The tone of the letter lacks something in self-composure or self-confidence and may indicate the national unease when confronting foreign standards. Within a very few years, Downing lost his eagerness to measure up and was far gentler in dispensing his own recommendations for beautifying American homes.

BOTANIC Nurseries, New Burgh, New York, Nov. 21, 1838.—I have just received the last Numbers of the *Arboretum Britannicum* and hardly know how to express my admiration of its completeness and magnificence, as a history of the trees and shrubs of temperate climates. I only regret that the high price at which so costly a work must be sold will prevent its having that general circulation here, and that deserved popularity, which your *Encyclopaedias* have found among us. Your *Encyclopaedias of Gardening* and of *Agriculture* are not only the standard works here, but they are almost exclusively the works found among our amateurs and better class of farmers and proprietors. The *Encyclopaedia of Cottage, Farm, and Villa Architecture,* although rather too elaborate to suit *our* popular taste, has already had a very visible effect upon the taste for rural architecture in the United States: and although we build up many edifices that set criticism at defiance, yet a wonderful progress and improvement in architecture has taken place with a few years past; much more, doubtless, than would be brought about in Europe in half a century; and our citizens and landholders only require good specimens, plans, and models, to adopt them at once, as we have no national or ancient prejudices of any sort to combat.

The facility with which we raise good fruit in the open air, in the Middle States, gives a great spur to the planting of fine fruits, and our nurseries contain nearly all the very choicest varieties cultivated at present with you: while such is the luxuriance of the soil, and so favourable is the climate, that numbers of fine seedling varieties spring up almost spontaneously. Some of the old fruits . . . bear and thrive in the Middle States yet, with all their primitive vigour. The *Magazine of Horticulture,* edited by Hovey at Boston, is slowly, though surely, labouring for the good of the cause among us. . . . In pretty villas in a high state of keeping, a fondness for rare plants and forcing the better fruits, Boston is half a century in advance of her sister cities. Philadelphia still holds the palm for fine exotic collections, and a general greenhouse commercial business. New York is so purely a business emporium, that in its pell-mell few find time for the indulgence of a taste for gardening: but some beautiful conservatories and suburban gardens have recently been erected in Brooklyn.

Judge Buel's excellent monthly paper, the *Cultivator,* is working wonders among our agricultural population, which is sadly in need of enlightening. . . .

A most complete and thoroughly scientific work on our North

American botany is in progress of publication, the first two numbers having been issued. Professors Torrey and Gray are the able
and distinguished authors; and the performance will tend greatly
to increase their reputation.

A. J. DOWNING

Americans had innovations of their own, in farmers' magazines designed to serve new groups settling onto western sections of land and
holdings remote from each other. "Family" magazines abounded. The
printed word nourished like grass and was nearly as obtainable, after
each little settlement set itself up with a church, school, store, inn, and
printing press. From these little presses we can still find handbills, pre-
cursors of advertising, that announced the advents of shipments of
seeds or the dates itinerant preachers would arrive to marry people
and preach while their mounts, stallions with brave names, improved
the breeding of local mares. Sometimes political tirades and lawsuits
seemed to sear the pages, but there was always something constructive
on pests or new sorts of apples available and ready for planting.

These publications had seemed "slip-slop" to Mrs. Trollope, though
she applauded the urge to print. She deplored, however, the idea that
anyone at all could write for these papers and then be read. So we can
imagine the impact of *Loudon's Encyclopaedia of Gardening*.

We are using the "new edition" of the *Encyclopaedia,* published in
1871, as edited by Mrs. Loudon. Loudon prefaced his 1834 edition by
saying that its object was to present

> in one systematic view, the History and present state of Gardening
> in all Countries, and its Theory and Practice in Great Britain.
> Under the term Gardening we include Horticulture, or all that re-
> lates to the kitchen-garden and the orchard; Floriculture, or all
> that relates to the flower-garden, the shrubbery and the culture of
> flowers and ornamental shrubs and trees; Arboriculture, or the
> formation of useful and ornamental plantations, and the culture of
> the most valuable timber trees; and Landscape Gardening, or the
> art of laying out grounds.

Subsequent paragraphs point out the superiority of his arrangement
over other encyclopedic forms, in that his first part, the history, is in

chronological order and the statistical section is in order of affinity. The last two parts, "The Science of Gardening" and "The Art and Practice of Gardening in Britain," are dealt with, he says, by merely presenting "the opinions and practices of others," leaving the reader to generalize or to particularize them for himself. The alternative, Loudon says, would have been to present the opinions and practices of others as if they were the author's own, giving the results as the opinions and practices of the author. By his present method, "the young reader is thus induced to think for himself." Those are the views of Mr. Loudon. Readers like ourselves might think of a third way, but Mr. Loudon likes to lay down laws for reading and thinking and gardening, and if we are to play Snakes and Ladders with him in the garden we must know his rules.

As, we see by *her* preface, did Mrs. Loudon. The "*execution* of the work" (she italicizes) has been conducted on the same principles as those that guided Mr. Loudon—inducing the young gardener to think for himself by giving the man of experience a choice of practices. She has carefully brought the history down to the present, rewritten the science, and added, to the third section on gardening in Britain, enlarged landscape gardening and added woodcuts.

The section on North America—carefully constructed to set the American public on the learners' bench—is divided into six "subjects." Number one, treating of gardening "as an Art of Design and Taste," begins: "Landscape Gardening as practiced in the United States is on a comparatively limited scale; because in a country where all men have equal rights, and where every man, however humble, has a house and garden of his own, it is not likely there should be many large parks." It is to be hoped that eventually everywhere in every country there will be parks "formed by towns and villages, inhabitants or members. . . . to this end only are the gardens of the monarchs and magnates of Europe at all worth studying." This elevating thought is immediately followed by a quotation of a comment made by a visitor to America in the autumn of 1834. "Before landing in New York," reported Mr. James McNab, "the country appears to a stranger of a very dark and dismal hue from the quantity of pines and red cedars which clothes the more conspicuous prominences; but after landing, the whole, from the prevalence of fine trees and shrubs, appears like one vast garden." Forest trees of large size are to be seen, covered to their summits with wild vines. He lists those preeminent: *Platanus occidentalis, Liriodendron, Liquidambar, Gleditsia triacanthos,* and the *Catalpa.* The only foreign trees, he said, "in the artificial scenery"—by which he means gardens—were fruit trees, the Lombardy poplar, and the weeping willow. Returning to the wilds, he notes that about sixty-seven miles upcountry on the river Hudson, in a limestone district, "the arbor-vitae succeeds to the dark hue of the red cedar."

From there on, to give the American public a sense of belonging and a reason for involvement, Loudon's *Encyclopaedia* follows with evaluations of gardens, parks, estates, and cemeteries, by individuals who had written books or articles from which these excerpts were taken with due acknowledgment. Hyde Park, the property of "the late Dr. Hosack," was observed through the views of four travelers. One of them considered it "first in point of landscape gardening in America," observing that the

> natural capacity of this seat for improvement has been taken advantage of in a very judicial manner . . . every circumstance . . . to beautify or adorn it. The mansion is splendid and convenient . . . park extensive . . . rides numerous . . . variety of delightful distant views embrace every kind of scenery. The pleasure-grounds are laid out on just principles . . . excellent range of hothouses . . . collection of rare plants.

Another visitor, Mrs. Trollope, considered the seat "magnificent," and a third declared it "quite a show place, in the English sense of the word."

WE have seen part of Downing's letter to Loudon in 1838, expressing his indebtedness to the English master on all matters related to landscape gardening. Downing had obviously read everything that Loudon had written.[1] We shall confine ourselves here to those of Loudon's writings most influential on the American scene: first, his *Gardener's Magazine,* to which both the Pennsylvania and Massachusetts horticultural societies made initial subscriptions for the benefit of their members; second, his *Encyclopaedia of Gardening,*[2] the sections dealing directly with contemporary accounts of American gardens; third, and important to us today, Loudon's *Villa Gardener,*[3] a succinct volume for small landholders in both city and country. These three lay down his "General Principles," which were inspirations for Downing, although Downing modified them for American readers.

Because our concern is to see how rules of procedure that crossed the Atlantic were tempered to suit New World gardens, we summarize here the laws laid down by John Claudius Loudon.

The edition of his *Encyclopaedia* published in 1871 announced that landscape gardening (the art of creating beautiful scenery out of plain, unornamented ground) is the highest branch of gardening. It can be defined as "the art of arranging the different parts which compose the external scenery of a country residence so as to produce the different beauties and conveniences of which that scene of domestic life is susceptible."

These *beauties* and *conveniences,* Loudon said, underlining as he went, depend upon differences in society and climate (and how true that could be on different sides of the Atlantic). Both are "objects to be desired by a wealthy man in his country residence." *Adaptation* from the common scenery of the country decides matters of *taste* and beauty. For example, when England was a wild country, the formal geometric garden was preferred; but as the countryside—the "general face of the country"—became highly cultivated and, with lanes and hedgerows, "disposed in artificial forms," the modern, "natural" style began to appeal. However, Loudon stated, both the geometrical and the natural styles have merit. "The principles of landscape gardening," he said, "like those of every other art, are founded on the end in view."

Before further clarifying his own principles, Loudon gathered in the views of his predecessors. He referred to Lord Kames, an eighteenth-century writer on aesthetics, who held that, because gardens and buildings may be destined for either use or beauty, or for both, the "differences or wavering of taste in these arts is greater than in any art that has a single destination."[4] Contrasting the naturally occurring asymmetry to ordered designs, Loudon noted that regularity and symmetry depend upon parts that correspond to each other. As Alexander Pope (who was quoted even by such relatively open-minded gardeners as Abigail Adams) expressed this interrelationship, "Grove nods at grove. Each alley has a brother, / And half the platform, Just reflects the other." Resisting the rigidly geometric, Loudon reduced Pope's

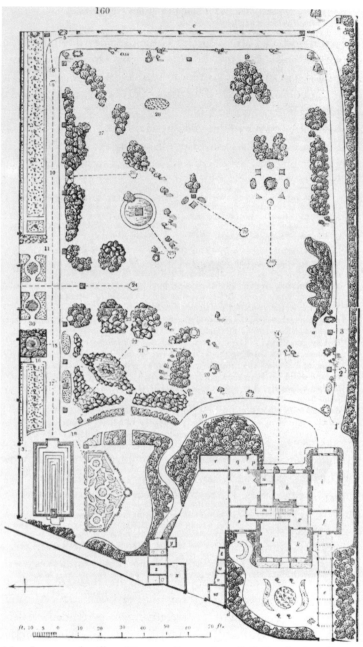

Mrs. Lawrence's villa at Drayton Green, from J. C. Loudon's *Villa Gardener,* the 1850 London edition, edited by Mrs. Loudon.

Garden features as we enter Mrs. Lawrence's garden from her house: an ivy vase, a basket containing a pyramid of roses, and an elevated rustic basket of pelargoniums. Grouped on one side are "a great many fine hybrid rhododendrons and azaleas. All these plantations and groups are treated in the picturesque manner; there being scarcely anything in these grounds, except the single plants, such as the standard roses, and some rhododendrons and other shrubs, which can be considered as treated in the gardenesque style of culture."

At the termination of the "Italian Walk," "the rustic arch and vase" from Mrs. Lawrence's award-winning garden.

The "span-roofed green-house."

The Italian walk, where a border on the left is planted with the most choice herbaceous flowers and standard roses.

Loudon's four examples of different landscaping techniques (his is "garden-esque"). From *The Modern and Approaching Style of Rural Improvement*. All reproduced by courtesy of Essex Institute, Salem, Massachusetts.

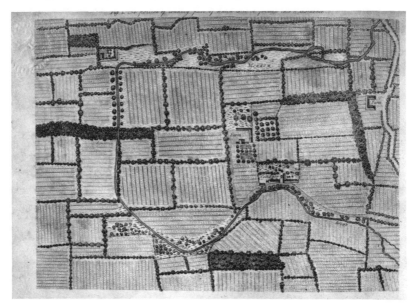

1. "A Portion of Country part of which is to be formed into a residence."

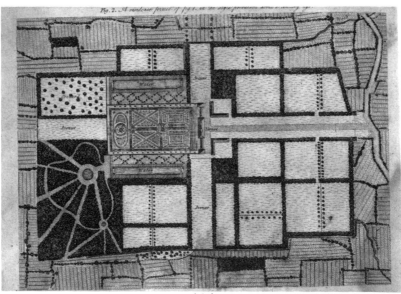

2. "A residence formed of Fig. 1 in the style prevalent a century ago."

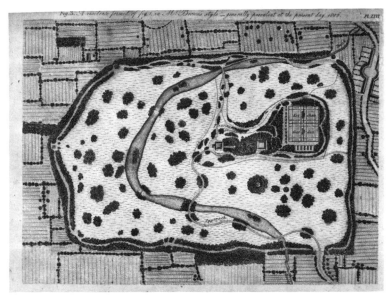

3. A residence formed of Fig. 1 in Mr. Brown's style—generally prevalent at the present day—1806": What Capability Brown would have done with this portion of country.

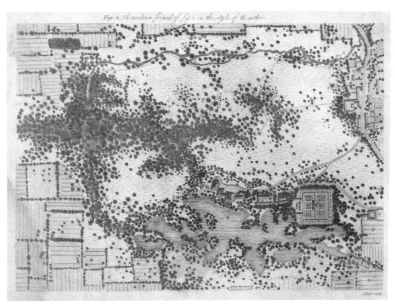

4. "A residence formed of Fig. 1 in the style of the author"—i.e., "gardenesque."

concern with garden design to three dry principles: "1st—To study and display natural beauties; 2nd—To conceal defects; and 3rd—Never to lose sight of common sense."[5]

This was followed by a very long quote from Whately's *Observations on Modern Gardening,* in which Whately "concurs" with Pope and states that the business of a landscape gardener is "to select and to apply whatever is great, elegant, or characteristic"; to "discover and show all the advantages of the place upon which he is employed; to supply its defects, to correct its faults, and to improve its beauties." Whately, Loudon noted with pleasure, considered landscape gardening "superior to landscape painting."

Continuing to draw upon his host of witnesses, Loudon commented that Repton emphasized the importance, for landscape gardening, of *congruity, utility, order, symmetry, scape, proportion,* and *appropriation.* The poet George Mason invoked simplicity, sensing a "[n]ice distinction beteen contrast and incongruity." Girardin included every beauty under Truth and Nature—"every rule under the unity of the whole and the connection of the parts." Shenstone (who coined the combination of landscape and gardening) held that "landscape or picturesque gardening" consisted in "pleasing the imagination" by scenes of *grandeur, beauty,* and *variety.* Both Price and Knight dealt with the principles of *congruity* in relation to painting and to gardening. Marshall favored *utility* and *taste* in imitating nature.

Having acknowledged these varying precedents, Loudon established his own definitions and rules. To some of us, although never to Mrs. Loudon, he may seem to be making words mean what he wants them to, but we must respectfully learn his terms. We may condense him, however, as he did those who came before.

Loudon relegated the ancient, geometric style to "mixed art," depending upon design guided by regularity and uniformity combined with obvious man-made works of art, like canals, straight avenues, trees planted in quincunxes, and so on. "Inventive art," on the other hand, is guided by natural and universal beauty, with designs intensifying the beauty of natural objects. Even objects not beautiful in themselves may become so when combined to form a whole—"one agreeable composite sensation." Modern landscape gardening is to a certain extent "an art of imitation" in that it results in "scenery composed of natural objects combined according to the rules of art."

Design, Loudon said, is of first importance in modern gardening, although it has no relation to the "primitive admiration for regularity" evinced in the geometric style. After design comes *fitness,* or the proper adaptation of means to an end.

In summary, landscape gardening considered as an imitative art has, as its chief object, the production of natural or universal beauty requiring imagination, feeling, and taste. Although some feel a criterion is that a landscape "if painted would form a tolerable picture," it must be remembered that landscape gardening deals with materials that are always changing.[6]

Loudon's principles come through clearly in *The Villa Gardener,* and here we find the definition of that gardening style upon which Loudon left his special stamp—the "gardenesque."

There are, said *The Villa Gardener,* certain principles common to architecture and gardening as fine arts: *design, taste,* and *fitness of the means employed.* A "fine art" is defined as "the creation, or composition, intended, through the eye or the ear to please the mind." Its two essentials are to create and to please. "Any creation to be recognized as a work of art must be such as can never be mistaken for a work of nature."

Loudon maintained that imitation of natural scenery by the hand of man may be rendered artistic in several ways. The first is by the use of native materials differently arranged. The second depends upon the employment of trees and shrubs totally different from those of the surrounding country.

He applied his definition to various garden components. *Ground* is to be reduced to levels or slopes of regular curvatures as in the ancient style or, following the modern style, to polished curvatures and undulations—each distinguishable at first sight from the natural surface of the ground. *Woods* must be treated differently if they are based on common local trees than if exotic foreign ones are used. If local trees are employed, they must be planted geometrically, in straight lines. Foreign trees, by contrast, must be in irregular masses or groups or placed as single trees. If indigenous trees are introduced, they must never be grouped as if naturally occurring but must be placed to stand distinct from others and to take forms more perfectly developed than in their accidental states. *Water* in imitations of natural scenery comes in two forms: running and still. The first resembles an existing brook,

Loudon's solution for living rationally in the country: Cut down the trees . . .
and substitute a larger flowerbed outside the drawing room windows. Both
from Mrs. Loudon's *Ladies' Country Companion, or How to Enjoy a Country
Life Rationally.*

the second an excavation to be filled with water. Both are to be planted with exotic material. Finally, *rocks* must not be outlandish and different but should be cut from local stone to make improvements—paths and the like. Art is easily recognized in all walks and roads by uniformity of breadth and evenness of surface.

The woods around a house must not be regularly planted, as in the geometric ideal of the past, but must be made to seem a part of an arrangement. Trees in a park must grow on heights as well as on plains. Water extent can be indicated by the sizes of nearby trees. Though *regularity* and *symmetry* were governing principles of the ancient style, *variety* was seldom attempted, except in flower beds, because it was prevalent in the surrounding country, before the advent of geometrically patterned fields. Variety in planting, desirable now, can be achieved by varying the dispositions of trees; groups of different sorts—as, for example, oaks and conifers—can be joined by mingling them gradually. Views from a house can be varied by an assortment of trees, shrubs, and flowers. Architectural ornaments furnish sources for varying views from a small place, providing all is combined with a *sense of order*. *Harmony* is produced by the introduction of details (vases, statues, rocks, and so on) that will connect the house with its grounds.

Loudon, having arranged all his tools and materials to be instantly convenient, now narrows his field to point up the features in which he is most interested. He has made clear in his earlier book that the two principal styles of laying out grounds in Great Britain are geometric and natural. He now dismisses the first as passé. He now divides the second into three variations: picturesque, gardenesque, and rustic. *Picturesque gardening* produces scenery for country residences that is "considered as particularly suitable for being represented by painting." It is an imitation of nature in its wild state. The *gardenesque* style is an imitation of nature "subjected to a certain degree of cultivation or improvement, suitable to the wants and wishes of man." It produces "that kind of scenery best calculated to display the individual beauty of trees, shrubs, and plants in a state of culture; the smoothness and greenness of lawns; and the smooth surfaces, curved directions, dryness and firmness of gravel walks; in short, the gardenesque style is calculated for displaying the art of a gardener."[7] The *rustic* style is

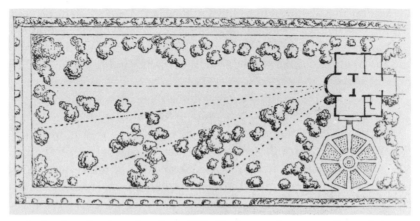

A planting plan from Loudon's *Villa Gardener,* to leave open lines of view
into the garden from the principal rooms of the house.

what is "commonly found accompanying the rudest description of labourers' cottages in the country."

Loudon insists that a planner must be familiar with these variations,
for normally the picturesque is not understood by gardeners, whereas
the gardenesque approach is foreign to professional landscape gardeners. Loudon offers examples illustrating the three aesthetics. To
make a picturesque imitation, he says we must have the eye of a landscape painter, the science of an architect and a botanist, and the knowledge of a horticulturist. If we take an old gravel pit, covered with indigenous trees and shrubs, with a hovel or rude cottage in the bottom
and with worn paths, and we introduce foreign trees, grass, a Swiss
cottage, and graveled walks, the result will "constitute a picturesque
imitation of natural scenery." On the other hand, where "the gardenesque style of imitating nature is to be employed, the trees, shrubs and
herbaceous plants must be separated." Trees must not touch each other
but must be placed near each other in varying degrees. In forest scenery, trees and shrubs may seem to spring from the same root, accompanied by large rampant herbs. The third style, the rustic, must seem
fortuitous, or as if produced by country laborers. A clay pit in this
style can be made into a seeming wild garden with walks and a cottage
seeming "half-ruined without." He noted that this approach is beloved
by people with a romantic turn of mind.

To the factors of garden planning noted in his earlier work, Loudon now added a new concern, *progress,* a "principle peculiarly characteristic of landscape gardening; providing for growth and also decay."

ANDREW JACKSON DOWNING accepted Loudon's principles respectfully, as chosen but not as carved. We shall see how he modified them for the settlers of the new nation.[8]

Born in 1815 into a family of nurserymen in Newburgh on the Hudson, Downing grew up to be a gentle but positive American dictator on gardening matters. Far younger than his brothers and never particularly sturdy, Downing was sent to the best school available and allowed to wander the countryside. He explored the rural landscapes, as well as the area's estate gardens, which belonged to gentlemen with extensive libraries, men who were able to indulge their inclinations in horticulture and botany. Under their guidance, Downing became widely read and vastly stimulated. He absorbed their manners and, at twenty-three, married the daughter of one of these households. He began to build a grand house in a somewhat Elizabethan style, reflective of his newly acquired tastes, on land purchased from his brother. In 1841, at the age of twenty-six, he published his first book, *Landscape Gardening: A treatise on the theory and practice of landscape gardening, adapted to North America; with a view to the improvement of country residences.* Our quotations come from a copy of the fourth edition, published in 1849 with additions "especially in that section relating to the nature of the Beautiful and the Picturesque," because "the difference among critics regarding natural expression and its reproduction in Landscape Gardening" had led Downing "more carefully to examine this part of the subject, in order, if possible to present it in the clearest and most definite manner." Authoritatively, he credited Loudon clearly.

Mr. Loudon's writings and labors in tasteful gardening are too well known to render it necessary that we should do more than allude to them here. Much of what is known of the art in this country undoubtedly is more or less directly to be referred to the influence of his published works. Although he is, as it seems to us, somewhat deficient as an artist in imagination, no previous author ever deduced, so clearly, sound artistical principles in Landscape Gardening and Rural Architecture; and fitness, good sense, and beauty are combined with much unity of feeling in all his works.

As Loudon had done with English examples, Downing examined existing American gardens and their particular beauties before proceeding to aesthetic theory. He declared that

consideration of different results to be sought after is vital, as what kinds of beauty we may hope to produce by Landscape Gardening. We must know the capacities of the art and about the schools or modes by which it has previously been characterized in order to judge our production as a work of taste and imagination.

Downing identified the beauties of the ancient style as regularity, symmetry, and the display of labored art. He noted that geometrical forms of buildings were carried into gardens. He considered this formal routine most suitable for public squares and public gardens, though it could serve well also in very small private gardens, with symmetrical and knotted beds, pleached⁹ alleys, and sheared trees.

According to Downing, men of genius saw the superiority of a more natural manner, and landscape gardening was raised to the rank of a fine art when the value of this approach was recognized. The earliest professors of landscape gardening, Downing felt, agreed upon two variations: *beautiful* and *picturesque*. The beautiful is characterized by simple, flowing forms; the picturesque "by striking, irregular, spirited" forms. *Beauty* obeys the universal laws of perfect existence, easily, freely, harmoniously, and without the display of power. The *picturesque* obeys the same laws rudely, violently, often displaying power only.

Downing proceeded to develop these two styles in landscape gardening. To illustrate scenery "distinctly expressive of each of these kinds of beauty," he described two scenes and compared them to different styles of painting.

The first scene is an undulating plain, covered with emerald turf, encompassed by rich outlines of a forest canopy, with noble groups of round-headed trees here and there interspersed with single specimens, with foliage drooping in masses to the turf beneath them. An azure heaven is reflected in a sylvan lake, with banks covered by flower-sprinkled turf or masses of verdant shrubs. "Here are all the elements of what is termed natural beauty—or a landscape characterized by

Downing's clarification of his idea of the two choices for American gardeners: the beautiful (above) and the picturesque (below). From Downing's "Historical Notices" section of *Landscape Gardening* (1859).

"Example of the Beautiful . . ."

"Example of the Picturesque . . ."

simple, easy and flowing lines." He compares this vision to the paintings of Claude Lorrain: graceful, flowing, and harmonious.

The second scene is in a nearby woody glen, perhaps a romantic valley, with steep rocky banks overhung by clustering vines and tangled thickets of deep foliage. Against the sky is the outline of an old, half-decayed tree, or the strongly marked forms of a horizontally branched larch or pine. Prominent in the foreground are rough and irregular trunks, or open glades of bright verdure opposed to dark masses of shadowy foliage. If there is water, it will be a cascade leaping over a rocky barrier or the murmur of a noisy brook. With an old mill in the middle ground, its wheel turned by the stream, we have a striking example of the picturesque. Downing refers us to the paintings of Salvator Rosa: romantic, bold, and vigorous.

These two examples will help us discriminate between the two styles; thinking of them, we shall be able to

> illustrate the difference in the expression of even single trees. . . .
> A few strongly marked objects, either picturesque or simply
> beautiful, will often confer character upon a whole landscape; as
> the destruction of a single group of bold rocks covered with wood
> may render a scene, once picturesque, completely insipid.

"By landscape gardening we understand not only an imitation in the grounds of a country residence, of the agreeable forms of nature, but an expressive, harmonious and refined imitation." We must "preserve only the spirit, or essence . . . and by heightening this expression . . . we may give our landscape gardens a higher charm than even the polish of art can bestow."

Downing quoted others on the beautiful and the picturesque: Repton insisted they are identical, Uvedale Price declared they are widely different, and William Gilpin defined picturesque as having "some quality capable of being illustrated in painting." Downing disagreed. Beauty, he said, "in all natural objects . . . arises from their expressions of those attributes of the Creator—infinity, unity, symmetry, proportion." Beauty in a living form is found when

> the individual is a harmonious and well balanced development of a
> fine type. In a beautiful human figure we see symmetry, propor-
> tion, unity and grace. In a beautiful tree, such as a fine American

elm, we see the perfect balance of all its parts resulting from growth under the most favorable influences . . . the finest form of a fine type.

But all nature is not equally beautiful. On all sides we see nature struggling with opposing forces. What is done only with violent and disturbed action—a struggle for the full expression of character, expressed with difficulty—we call the picturesque.

Combinations are possible, though not within a small compass. They are often combined in natural landscapes; within a small space, either one or the other must be aimed at. However, a landscape gardener who can combine the two, bringing them into harmony, has "reached the ultimate of his art."

Of course, there are other expressions in nature besides the beautiful and the picturesque—for instance, *grandeur* and *sublimity*. But these, at least in the comparatively limited scale of the landscape gardener's operations in this country, are almost wholly beyond his powers. Where they exist in the natural landscape (as they often do in this country), the landscape gardener has only to make his art "accord with, or at least not violate, the higher and predominant expression of the whole."

Certain qualities of the beautiful can be so elicited or created by art as to give distinct characters to small country residences or to portions of larger ones. "These are: simplicity, dignity, grace, elegance, gaiety, chasteness, etc." For instance, a few fine trees scattered or grouped over a smooth lawn will give a character of simple beauty. Lofty trees of great age, an elevation commanding a wide view, and a hill covered with rich wood stamp a site with dignity. Trees of full and graceful habit, or gently curving forms in the lawn and walks, convey the idea of grace. Tall trees of rare species, or an abundance of flowering shrubs and plants, confer characters of elegance and gaiety.

Downing describes the beautiful in general, practical terms, as applied to anyone's property: The outlines must be curved and gradual, surfaces soft and luxuriant, ground easily undulating, trees with full symmetrical heads or luxuriant, drooping branches grouped to allow free development, walks following the shapes of the surfaces, water in sheets with curved margins (embellished by full masses of shrubs and flowing outlines of trees) or winding in streams. All must be kept in polished order and neatness. The trees and shrubs should be of the finest foreign sorts. Flowering plants are to be arranged "in the most dressed portions near the house."

And here we may find ourselves startled, having, perhaps, not planned on going beyond gardens. Downing declared that the house itself in such a landscape should properly belong to one of the classical modes (the Italian, Tuscan, or Venetian, preferably), because these lend themselves to "the graceful accompaniments of vases, urns and other harmonious accessories." A plainer dwelling in such a setting should have "its veranda festooned with the finest climbers."

When Downing sets us down in a picturesquely landscaped garden, the outlines have a spirited irregularity, surfaces are abrupt and broken, and growth is of a wild, bold character. The ground has sudden variations in its occasionally smooth parts, with dingles, rocks, and broken banks. Trees are old and irregular, with pines, larches, and other trees of strikingly asymmetrical growth in groups with every variety of form. Walks and roads are abrupt. Water as wild as in nature invades, wood-fringed or cascading. The keeping here is less careful than in the graceful school, the lawn less frequently mown, edges less carefully trimmed. And the style of house? Gothic mansion or an Old English or Swiss cottage. Rustic baskets may abound.

Downing endeavored to play fairly with each style. Yet he questioned why, if he declares the beautiful the more perfect, the picturesque is "so much more attractive to many minds." He also supplied the answer. He conceived that it is due

> partly to the imperfection of our natures by which most of us sympathize more with that in which the struggle between spirit and matter is most apparent, than with that in which the union is harmonious and complete. . . . The manifestation of power is to many minds far more captivating than that of beauty.

Observing his American countrymen, Downing analyzed the three classes of people who "enjoy the charms of Landscape Gardening": those who have arrived only at a primitive idea of beauty as found in regular forms and straight lines; those who seek for the highest and most perfect development of beauty in material form; and those who enjoy in the picturesque a wild and incomplete harmony between the idea and the forms in which it is expressed.

In America, said Downing (and we are about to apply all we have

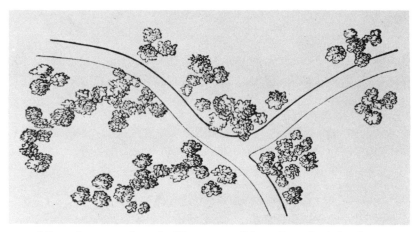

"Grouping to produce the Picturesque," from Downing's *Landscape Gardening*.

learned), there are always circumstances controlling an amateur's choice between the beautiful and the picturesque—for instance, fixed locality, expense, and individual preference as to style of building. In the older parts of the country, a great variety of attractive sites afford opportunity for either taste. Lately, Downing had found the picturesque in ascendance, because all American rivers and brooks, woods, and terrain can be so readily appropriated with great effect and with little art necessary. "In the picturesque mode the annual tax on the purse . . . is so comparatively little, and the charm so great."

However, the larger demesne with its swelling hills and noble masses of wood ("may we not, prospectively, say the rolling prairie too?") should always, in the hands of the man of wealth, be made to display all the breadth, variety, and harmony of both the beautiful and the picturesque.

And what of small cottage places? How shall their owners attempt to improve their grounds, with not enough time, room, or income to follow either school? By "attempting only the simple and natural," the unfailing way being to employ only a soft verdant lawn, a few well-grouped forest or ornamental trees, walks, and some flowers—enduring sources of enjoyment.

Certain universal and inherent beauties are common to all styles: *unity, harmony,* and *variety.* Unity takes highest importance, as the *pro-*

"English Rural Cottage," from *The Horticulturist,* vol. 2, under Downing's editorship.

duction of a whole. Its violations are indicative of the absence of correct artistic taste: a view of natural groups of trees and shrubs with a formal avenue leading to the house; fruit trees among ornamental trees on a lawn; vegetables and flowers in the same beds. Unity demands some grand or leading features in a view, to which others appear subordinate. Even in walks, some should be larger than others.

Variety is a fertile source of beauty, belonging more to details than to the whole. Different scenes, a thousand points of interest, different species of vegetation, many kinds of walks, ornamental objects, buildings and seats—all will keep alive the interest of the spectator.

Harmony keeps variety from becoming discordant. One must combine qualities in colors of foliage and in embellishments like sculptured vases and rustic seats (in keeping with the spirit of the scene).

Downing summarized the grand principles important to successful practice of this elegant art:

The Imitation of the Beauty of Expression derived from a refined perception of the sentiment of nature; the Recognition of Art,

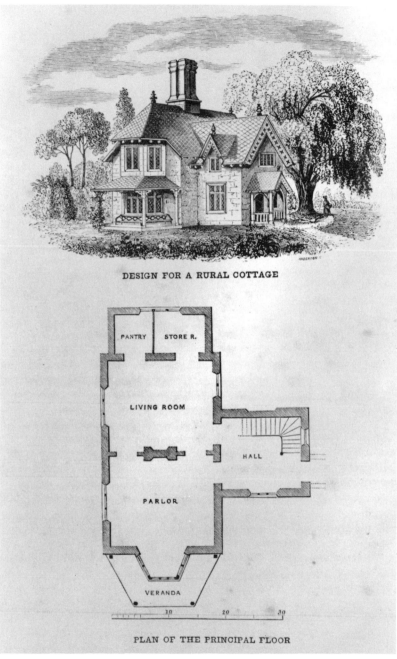

DESIGN FOR A RURAL COTTAGE

PANTRY   STORE R.

LIVING ROOM

HALL

PARLOR

VERANDA

PLAN OF THE PRINCIPAL FLOOR

Design for a Rural Cottage (American) from *The Horticulturist,* vol. 2,
under Downing's editorship. Note the somewhat unrealistic plan of
rooms. Did they have an outside kitchen?

founded on the immutability of the true as well as the beautiful; and the Production of Unity, Harmony and Variety in order to render complete and continuous our enjoyment of any artistical work.

NOTES TO CHAPTER 10

1 Loudon's four encyclopedias: *Encyclopaedia of Cottage, Farm, and Villa Architecture; Encyclopaedia of Gardening; Encyclopaedia of Plants;* and *Encyclopaedia of Trees and Shrubs,* an abridgment of his nine-volume *Arboretium et Fruticetum Britannicum* (adapted for "the use of Nurserymen, Gardeners and Foresters"), would be enough to immortalize several men. But there was an incredible amount of other writing and publishing as well.

2 First published in 1822. My reference copy is the "new edition" of 1871, edited by Mrs. Loudon.

3 We are working from the second edition, of 1850, also edited by Mrs. Loudon. Some horticultural sleuth could find a fascinating project in searching for the origins of Loudon's principles and then tracing them through the gentle if improving hands of the faithful Mrs. Loudon. Her dedication to persuading ladies to become educated gardeners became a cause shared by her American editor, Downing.

4 In his *Elements of Criticism,* a copy of which was in Jefferson's library.

5 Loudon did not include Pope's further admonitions to consult, as Virgil suggested, "the Genius of the Place," and to observe how "He gains all ends who pleasingly confounds, / Surprises, varies and conceals his bounds."

6 The preceding comments are drawn from Loudon's *Encyclopaedia of Gardening.*

7 Loudon also observed that it "is easier to form a perfect landscape gardener out of a gardener and botanist than out of a landscape-painter or an architect."

8 Downing began his preface to his *Fruits and Fruit Trees of America,* published in 1845, by saying, "A man born in one of the largest gardens and upon the banks of one of the noblest rivers in America ought to have a natural right to talk about fruit trees." Fruit, he says, is the "most perfect union of the useful and the beautiful that the earth knows."

9 Made of plaited or interlaced branches or twigs.

ELEVEN

Later Links

ROM Dr. Jared Kirtland we can get an idea of the influence of Downing and his *Horticulturist*. Writing in a Cleveland magazine called *The Family Visitor,* in the issue for July 11, 1850, Kirtland states that *The Horticulturist*

> has progressed to the completion of the fourth volume and is about to enter upon the fifth. To speak its merits would be superfluous as it is now universally known. Not a nook or corner of the West can be found that does not exhibit evidence that it has been affected by its influence. Every new tenement, every garden, lawn and group of ornamental and fruit trees shows about it some "Downing touch."
>
> We would urge it upon all of our friends, especially those who have an anxiety that "the young twig should be rightly bent" to introduce it as a part of their household reading. It is doing much to develop and cultivate a correct taste.

In 1850, the year of Kirtland's comment, Downing traveled to England and France for the first time. He was the acknowledged authority on American fruits. He had introduced his countrymen to the works of both Mr. and Mrs. Loudon. He was venturing into architecture.[1] Despite his ready quotes of foreign gardens and landscape philosophies, his assurance and authority on all things foreign, he had never been abroad. Most of what he knew and felt was self-taught. It was high time for him to go abroad, to see what was going on far from his suggestions or supervision. One of his reasons for traveling may have been to find an assistant, especially one skilled in building, to help him continue his efforts toward combining utility and beauty in American domestic architecture.

His visit to England could not have been more satisfactory. He was welcomed and made much of. He moved about as he pleased and wrote "letters" for *The Horticulturist* with just appraisals of many foreign sights and sites. He met a duchess who weeded as she walked and talked. And he brought back to America with him an able young architect, Calvert Vaux, the next link in our continuing chain.

This was fortunate, for Downing began to find himself much in demand, his country increasingly dependent upon his advice and inspiration. He was asked to lay out the grounds of government buildings in Washington—the classical Capitol and White House, the exuberantly Victorian Smithsonian Institution. He was engaged in private work on the Hudson, on Long Island, and in Newport. It was on a trip to Newport with his family and friends, in July 1852 when he was thirty-six years old, that Downing was involved in an accident on the Hudson between two racing steamers. He drowned trying to save friends who could not swim.

Downing is commemorated in Washington in those of his designs that were carried out after his death, and by an official memorial, a large Warwick-style vase on a plinth of a design that would have pleased Hovey. Downing's protégé, Vaux, catapulted by events into a position of importance as his successor, recorded it as a "large, white marble vase, modeled from a chaste but highly enriched antique example and mounted on an appropriate pedestal."

As a means of establishing the strength of the chain linking those most influential in nineteenth-century American landscape gardening, we present descriptions of two estates that were celebrated examples of the art of the times. Both are located on the banks of the Hudson. One is the domain of Andrew Jackson Downing himself. The other is a residence called Montgomery Place, which Downing admired and for which we have his description.

The leading article in *The Magazine of Horticulture* for November 1841 is announced as the first of a series on

> Select Villa Residences with Descriptive Notices of each; accompanied with Remarks and Observations on the Principles and practice of Landscape Gardening: intended with a view to illustrate the Art of Laying Out, Arranging, and Forming Gardens and Ornamental Grounds.

The capitalization is the editor's own. We do not know why "practice" seems less vital than "Principles" to the author and editor of the magazine, C. M. Hovey. The article, extremely long, begins:

> There are no situations in the country better adapted for beautiful residences, than are to be found on the banks of the majestic Hudson. Its whole course . . . one series of splendid and everchanging scenes . . . steep and abrupt declivities, clothed . . . with lofty pines and hemlocks, oaks, etc. which from the great height of its banks appear as mere shrubs . . . open and broad glades of rich and fertile country . . . backed by ranges of mountains . . . clustered dwellings which mark the numerous villages.

Hovey describes these villages (of which Downing's Newburgh is one) as lying ten or twelve miles above West Point, on banks from 50 to 150 feet above the Hudson, with their principal streets laid out in a grid parallel to the river. Hovey regrets to see that General Washington's headquarters look neglected, but he considers the residence of Mr. Downing well worth a detailed description. He presents a ground plan, with a carefully numbered list of features.

At the entrance, in the lower center of the plan, Hovey was disappointed with the Grecian-style gate, which he found not in keeping with the Gothic style of the house. The drive ran past the service buildings and past the formally laid-out flower garden in front of the house, from which a vast lawn descended gradually to the nursery grounds in the lower front (where care was taken not to interrupt the view of the river). Arabesque beds on the lawn contained "choice flowers, such as roses, geraniums, fuchsias, *Salvia patens, Salvia fulgens,* and *Salvia cardinalis,* to be turned out of pots in the summer season after being wintered in greenhouses or frames." Hovey warned that

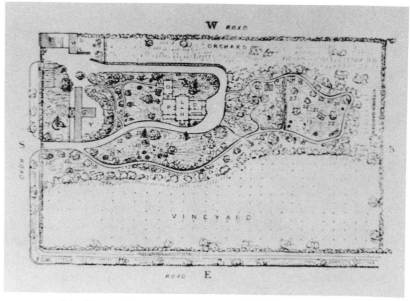

Mr. Downing's Residence: Plan of House and Grounds.

such beds should be sparingly introduced, or they would give the lawn a frittered appearance by cutting it up to an extent which would destroy its breadth, which constitutes its greatest beauty. It is even considered by some landscape writers rather an error to introduce any forms but the circle, unless the beds are looked down upon from an elevated terrace, when these arabesque shapes will have a pretty appearance.

The circular beds in the plan were for "petunias, verbenas (which now form one of the principal ornaments of the garden), *Phlox drummondi*, nemophilas, nolanas, dwarf morning-glory etc." The flower garden in front of the greenhouse was laid out in circular beds, edged with box and with gravel walks. An arborvitae hedge screened the back shed and the compost ground.

The sundial, at the lower point of the lawn just opposite the last arabesque on the right, received special notice from Hovey as "an old but suitable ornament" (which—as it so happened—could be obtained in Boston from Messrs Hovey & Co.). He notes that its "pedestal is executed in the Gothic style and we do not know of a single object which would add so much to the finished appearance of the lawn." There were several large palms in pots, "Maltese vases" (vases made of artificial stone), set in the turf. Hovey was anxious that vases be set down not on the turf but on a plinth. He noted that the *Agave americana,* especially the variety *variegata,* gives a good effect in such situations. Rustic baskets were also represented, and Hovey quoted Downing's directions for making them from his *Landscape Gardening:* "An octagon box serves as the body or frame of the vase; on this, pieces of birch and hazel (small split limbs covered with the bark) are nailed closely, so as to form a sort of mosaic covering to the whole exterior."

Hovey listed the specimen trees most suitably dotted about: *Thuja occidentalis;* a group of magnolias (*Magnolia conspicua, M. tripetala, M. auriculata, M. glauca* var *longiflora,* and *M. soulangiana*); a beautiful wahoo elm (*Ulmus alata*); a *Salisburia adiantifolia;* and a fine weeping ash (*Fraxinus excelsior* var.). Two sets of grouped trees deserved attention. One included the Osage orange (*Maclura aurantiaca*), the American linden (*Tilia americana*), weeping cherry (*Cerasus vulgaris* var.), and a *Virgilia lutea.* The other was composed of *Sophora japonica, Acer striatum,* and *Acer negundo.*

Hovey then took his reader on a tour of the house, where he noticed a full set of Loudon's works in the library, and then out to admire the view. Downing's house was designed by its owner and built, under his supervision, about two years before this visit (which explains the small sizes of the trees on the lawn). Hovey described the extensive nursery, the plants of which will be described in the Appendix, and ended by saying he believed Downing was

> the only person at present in the country who is consulted professionally as a landscape gardener. We are glad to learn from Mr. Downing that his *Treatise* on the subject has succeeded beyond his expectations, and we hope it will be the means of awakening the public taste to the importance of the art.

Downing began his account of Montgomery Place in *The Horticulturist* for October 1847 by saying that there

> are few persons, among what may be called the traveling class, who know the beauty of the finest American country seats . . . in

Downing's residence at Newburgh on the Hudson.

The library at Downing's residence.

the older and wealthier parts of the country . . . landscape gardening beauties that are developing themselves every day with the advancing prosperity of the country. The annual tourist by the railroad and steamboat, who moves through wood and meadow and river and hill with the celerity of a rocket . . . is in a state of total ignorance of their many attractions.[2]

He admitted, however, that these places were not open to strangers. It is due to his own good fortune that he was able to know the finest of these delightful residences and to "chronicle their charms."

About a hundred miles from New York (only six hours by steamer), Downing tells us there is a twenty-mile stretch on the eastern shore of the Hudson with a nearly continuous succession of fine seats. Surrounded by extensive pleasure grounds, fine woods, or parks, the adjoining estates are often concealed from each other, with the broad Hudson forming the grand feature in all their varied landscapes.

The estate under consideration was named for General Montgomery, hero and martyr of Quebec, whose widow resided here for the latter part of her life and then left the estate to her brother, Edward Liv-

Montgomery Place. From A. J. Downing, *Rural Essays* (1881).

ingston, American minister to France, whose family was in residence as Downing wrote. Downing's account is full of personal musings, French quotations (from, we think, the Abbé Rapin), English verses by Rogers, and bows to the author of *Rip Van Winkle*.

He conducts us on a tour of the 400 acres—with its many miles of walks and drives—by taking us first to the entrance on the post road, east of the property. The Hudson lies to the west; on the north lies the hidden residence of Blithewood; the south is bounded by a vast wooded area containing a "morning walk" and a "wilderness." The tree-planted drive, Downing says, sweeps like the entrance drive to an old French château and, halfway to the mansion, it runs into a tall wood, which is immediately succeeded by the lawn, "with increased effect after the deeper shadows of this vestibule-like wood." Specimen trees (hemlock, lime, ash, and fir) form the finest possible accessories to the spacious mansion, "one of the best specimens of our manor houses." The "pavilion," thirty feet wide, formed the north wing, in the style of an Italian arcade. The large lawn is "varied with fine groups, and margined with rich belts of foliage." The view, described at every hour of the day, has islands in the foreground and mountains in the distance.

Toward the left of the housefront one can start out on the "morning walk." Fortunately for the weary, this is laid out with frequent rustic-gabled seats, one of which has a moss roof and some of which have rustic railings edging the eighty-foot drop to the river. Stone steps lead us back up to the lawn, where there is "a rustic seat with a thatched canopy, curiously built round the trunk of an aged pine." The walk, however, continues, descending rapidly again toward the river to another rustic pavilion, on a jutting point of land, where a boat lies ready for an excursion. Foregoing this pleasure, we enter the "wilderness." But this is not at all savage. It contains "evidences of every 'improvement'" possible resulting from "a fine appreciation of the natural charms of the locality."

> The whole of this richly wooded valley is threaded with walks, ingeniously and naturally conducted so as to penetrate to all the most interesting points; while a great variety of rustic seats, formed beneath the trees, in deep secluded thickets, by the side of the swift rushing stream, or on some inviting eminence, enables one fully to enjoy them.

After pursuing several miles of walks and hearing a waterfall, the visitor emerges for a full view of the first cataract. Stone steps rise forty feet beside it, to the second cataract. At the top lies "the lake," with a rustic temple. And the traveler from here makes a return trip by way of the conservatory.

Near the conservatory, a "large, isolated, glazed structure," one passes under "tasteful archways of wire-work" covered with rare climbers and enters "what is properly the flower garden."

> Here all is gay and smiling. Bright parterres of brilliant flowers bask in the full daylight and rich masses of colours seem to revel in the sunshine . . . walks fancifully laid out so as to form a tasteful whole . . . beds surrounded by low edgings of turf or box, and the whole looks like some rich oriental pattern or carpet of embroidery. In the centre . . . a large vase of the Warwick pattern; others occupy the centres of parterres in the midst of its two main divisions and at either end is a fanciful light summerhouse or pavilion, of Moresque character. The whole garden is surrounded and shut out from the lawn by a belt of shrubbery, and, above and behind this, rises, like a noble framework, the background of trees of the lawn and the Wilderness. If there is any prettier flower-garden scene than this *ensemble* in the country, we have not yet had the good fortune to behold it.

One more feature must be described. On the southern boundary lies a wood of about fifty acres, nearly level or undulating in surface, well covered with oak, chestnut, and other timber trees. Laid out through it is "the Drive—a sylvan route as agreeable for exercise in the carriage, or on horseback, as the 'Wilderness' or 'Morning Walk' is for a ramble on foot." No small additional charm to a country place, in the eyes of many people, is this "secluded and perfectly private drive, entirely within its own limits."

In summary, Downing notes that, though

> Montgomery Place itself is old, yet a spirit ever new directs the improvements . . . an arboretum just commenced on a fine site in the pleasure grounds . . . natural beauty . . . elicited and heightened everywhere in a tasteful and judicious manner . . . numberless lessons for the landscape gardener . . . an hundred points that

Montgomery Place: The Lake. From Downing, *Rural Essays*.

will delight the artist . . . a thousand aspects of nature for the poet; and the man of the world, engaged in a feverish pursuit of its gold and its glitter, may here taste something of the beauty and refinement of rural life in its highest aspect.

So we come to the fourth link in our chain, Calvert Vaux, who was twenty-five when Downing met him in London and hired him. They left England together, and Vaux became Downing's partner in preparing the plans for Washington.

Vaux wrote his only book after Downing's death. His *Villas and Cottages: A Series of Designs Prepared for Execution in the United States* was published in 1857, the year he moved to New York.[3] The book was dedicated to "Caroline E. Downing and to the memory of her husband Andrew J. Downing." In his two prefaces (the second coming from the edition of 1864), Vaux demonstrated that he had indeed caught the torch from Downing.

In 1857, Vaux wrote:

Every American who is in the habit of traveling, which is almost . . . every American, must have noticed the inexhaustible demand

Montgomery Place: The Conservatory. From Downing, *Rural Essays*.

for rural residences that is perceptible in every part of these Northern States. . . . [H]ard times undoubtedly occur . . . [but] the
carpenters and masons appear to be always getting a full percent-
age of the floating capital. . . .

It cannot be possible that the energetic vitality which pervades
this branch of home manufacture . . . will remain satisfied to ex-
pend its intensity on meagre, monotonous, unartistic buildings
. . . without perceiving the propriety of getting habitually something worth having for the money. . . . ugly buildings should be
the exception, not, as hitherto, the almost invariable rule.

The accompanying designs . . . [are] not model designs to
lessen the necessity for the exercise of individual taste, but . . .
[are intended] to increase its activity . . . as stepping stones . . .
for comparison and criticism. . . .

In this collection . . . [are] many [engravings] marked "D. and
V." . . . the latest over which the genial influence of the lamented
Downing was exercised. . . . Several . . . were in progress . . .
when the tidings of his sudden and shocking death were mourn-
fully received. . . .

Andrew Jackson Downing was not only one of the most energetic and unprejudiced artists that have yet appeared in America,
but his views and aspirations were so liberal and pure that his artistic aspirations were chiefly valued by him as handmaids to his
higher and diviner views of life and beauty. . . . [H]is character
being moulded on this large scale and his capacity to appreciate
whatever is beautiful in nature or art being proportionately great,
he had both the will and the power to exercise a marked influence
for good over the taste of his countrymen . . . [and] strove to
make manifest in all his works the glorious truth that the really
beautiful and the really good are one. . . .

"Il bello e il buono" was the motto engraved on his seal and on
his life.

By the time of the 1864 preface, the face of the country was obviously changing. In noting that "[e]ach of the European nations that
have contributed to the population of this country has distinctive peculiarities," Vaux showed that he belonged to the developing class of
Americans concerned with "the social orders," anxious that "the
American million" be given "the rudiments of self-reliant greatness."
Where not money but moneymaking is all-important, Vaux observed,
advancement in the arts had understandably lagged behind commercial and political progress. Because most money spent for building in
the United States was controlled by the industrious classes, the active
workers were in charge of the national standard of architectural taste.
Vaux was concerned with how these workmen might be given a reasonable chance to choose well.

His book offered thirty-nine designs. Some had already been built,
in rural New York and Rhode Island. Eight were marked "D. and V."
Vaux gave many architectural tips but said little about gardens.[4] So we
leave the artisan to be improved in taste and proceed to Vaux's contribution to Central Park.

NOTES TO CHAPTER 11

1   *The Fruits and Fruit Trees of America*, written in collaboration with his brother
Charles, was published in both New York and London in 1845 and went into thirteen
printings in Downing's lifetime (which is to say, only another seven years). In 1845 he

had also edited Mrs. Loudon's *Gardening for Ladies* and became editor of *The Horti-culturist*. In 1850 he brought out his *Architecture of Country Houses, Including Designs for Cottages, Farm Houses, and Villas*.

2   In volume 2, number 4, of the bound numbers of *The Horticulturist, and Journal of Rural Art and Rural Taste* (October 1847).

3   He stayed there for the rest of his life and died in 1895.

4   "Every rural building requires four tints to make it a pleasurable object in the way of color." A water closet is "an absolute necessity." Bay windows make desirable addi-tions. But the only garden-related ideas are two designs for shaded seats, which were "executed in Central Park, New York."

TWELVE

Greensward into Central Park

HE roots and origins of anything from plants to ideas run farther back than we need to go to realize their fruits. Credit for development of the natural, "picturesque" style in landscape design, however, belongs clearly to the English scholars, squires, artists, and gardeners of the end of the eighteenth century. The consideration of proper views of beauty, proportion, seemliness, emotional response, and classical heritage produced great literary output in leatherbound books and new magazines. It was not just that everyone was tired of gardens where one side had to mirror the other to be correct. The public's enthusiasm must have grown from the stimulating idea that nature was both a force and a blessing to be accommodated and celebrated in garden design.

At sixteen, young Frederick Law Olmsted, the next link in our chain, after roaming the Connecticut fields and woods and searching through the Hartford Public Library, was enthralled to come upon the ponderous dicta of Uvedale Price and William Gilpin on the "picturesque." Years later he would reread them with reference to laying out great open spaces for American public parks. For the most part, however, at the beginning of the nineteenth century in North America, the dissemination of practical advice on farming practices held first place in public attention.

When young Olmsted (about whom so much has been written lately that we confine ourselves here to what we find most useful and attractive in his long and exemplary life) was found ailing and unable to attend Yale College, his family decided to make him a farmer. Commerce and engineering had been tried. He had spent a year voyaging to China and back. Twice he settled down on farms his father purchased for him. In common with the times and with the views of his father and stepmother, he had become concerned with religion. The redeeming features of his youth were the planned, comfortably arranged carriage tours around New England with his family to the outstandingly beautiful landscapes of the countryside. Both his father and his stepmother were keenly aware of rural beauties, and these regular excursions were to stand the boy in great stead later in his life.[1]

Even after triumphant American experiments in adapting natural landscape design to cemeteries, the single idea of undertaking a great public park for New York City was like a flower in the buttonhole of a successful man celebrating his arrival in a competitive society. Mr. Harison, a lawyer of early nineteenth-century New York, attending the board meeting of his bank with his own new double yellow rosebud in his lapel, was a symbolic forerunner of his city's joining the great cities of the world by laying out a huge tract of land to be cultivated for the use of its citizens.[2] London, Liverpool, Paris, Berlin, Amsterdam—one could begin to count cities with public parks on more than one hand.

It was high time for Americans to join them when, in 1849, Downing pointed out the social and financial benefits of a place set aside for public enjoyment beyond that being currently sought in American cemeteries.

In 1850 Downing wrote of London parks in his "English letters" to *The Horticulturist*. He said he had concentrated his previous writings on the English countryside, because he was, as an American, "accustomed to the clear, pure, transatlantic atmosphere" and was appalled by the "taint of smoke" and the "black and dingy look of all buildings" in London. He desired now to make amends and to present the real lights of that "great world in itself," its "grand and beautiful parks." He noted that in "the midst of London lie, in almost connected series, all the great parks, Hyde Park, Regent's Park, St. James and Green Parks"—names as familiar to his readers as the Battery and

Washington Square. But, unlike these last two and unlike anything on the Continent,

> London parks are actually like districts of open country—meadows and fields . . . lakes and streams, gardens and shrubberies, with as much variety as if you were in the heart of Cambridgeshire and as much seclusion in some parts at certain hours as if you were on a farm in the interior of Pennsylvania.

Downing carefully takes his readers on a tour of all four parks, and ends his letter

> with only a single remark—that we fancy in New York that the introduction of Croton water is so marvelous a luxury in the way of health that nothing more need be done for the comfort of half a million people. . . . Is New York really not rich enough, or is there absolutely not land enough in America to give our citizens public parks of more than ten acres?

Also in 1850, Olmsted, vacationing from farming, visited England for a six-month walking tour with his brother and a friend. "Expenses of journey about $300.00 each. Two weeks in Germany, two weeks in Belgium and France, one week in Ireland, three weeks in Scotland, remainder in England." This experience culminated in a book, published in 1852 as *An American Farmer in England*. American farmers (among whom Olmsted counted himself a beginner) came out well in almost all comparisons, but he noted that Americans lagged in attention to the needs of the poor and underprivileged.

Birkenhead and its park attracted his special notice.

> Birkenhead is the most important suburb of Liverpool . . . same relation to it that Charlestown has to Boston or Brooklyn to New York. . . . [T]he very liberal and enterprising policy of the landowners affords an example that might be profitably followed in the vicinity of many of our own large towns . . . public squares . . . streets and places are broad . . . with reference to general effect . . . a square of eight or ten acres . . . tasteful masses of shrubbery and gravel walks. . . .

> But the revelation waited on a baker who begged us not to leave Birkenhead without seeing their *new park* . . . a mile and a half

from the ferry and quite back of the town . . . massive black gateway of Ionic architecture standing alone . . . heavy and awkward . . . a large archway for carriages and two smaller ones for those on foot . . . gates freely open to the public . . . a short distance into a thick, luxuriant and diversified garden . . . art employed to obtain from nature so much beauty I was ready to admit that in democratic America there was nothing to be thought of as comparable with this People's Garden. . . . Winding paths over acres and acres with a constantly varying surface, where on all sides were growing every variety of shrubs and flowers, with more than natural grace, all set in borders of greenest, closest turf and all kept with consummate neatness.

Eagerly venturing farther, the travelers came to an open field, a tent, and a cricket ground for men on one end and boys on the other. Beyond this was a large meadow with trees, where girls and women played with children. When a shower threatened, they sought shelter in a pagoda on an island approached by a Chinese bridge. A crowd gathered there, and the visitors observed that the privileges of the garden were enjoyed equally by all classes.

The land, they were told, had been a flat, clay farm until 1844, when it was placed in the hands of Mr. Paxton, who laid out its present form the following year.

Carriage roads, thirty-four feet wide, with borders of ten feet and walks of varying width . . . excavation for a pond . . . earth obtained from these sources used for making mounds and to vary the surface . . . with much *naturalness* and taste . . . whole ground thoroughly underdrained . . . to supply the lake . . . stocked with aquatic plants, goldfish and swans . . . stones on the ground not used . . . laid in masses of rockwork and mosses and rockplants attached . . . the mounds planted with shrubs and heaths and ferns and the beds with flowering plants . . . the turf . . . kept close cut . . . with scythes and shears and swept . . . rural lodges, temple, pavilion, bridges, orchestra for a band of instrumental music etc. . . .

And so in one year the skeleton of this delightful garden was complete. But this is but a small part. . . . besides the cricket and an archery ground, large valleys were made verdant . . . extensive drives arranged—plantations, clumps and avenues of trees formed and a large park laid out. And all this magnificent pleasure-ground is entirely unreservedly and forever the people's own.

This was the summer of 1850, when Downing was also in England searching for a young assistant, expert in architecture—whom he found in Calvert Vaux. In 1851, Olmsted, home again and ready to withdraw from farming to devote himself to writing and possibly publishing, visited Downing in Newburgh. Olmsted's book on England had been well received, and he had been asked to journey through the "Seaboard Slave States" and to write a similar account. By 1857, however, Olmsted was disillusioned with publishing as a life business. He was appointed superintendent of the proposed Central Park in New York. From that more or less assured eminence, he agreed to collaborate secretly with Vaux in an overall plan for the park.

The account of this venture reads like a popular success story of the times, sold in paperback to inspire ambitious young Americans. First, Downing writes of his dream, hires a young Englishman to assist him in his mission to beautify American homes, and then dies tragically young, in 1852. In New York that very year, the politicians and city fathers move to alleviate the city's congestion, finding the Battery suddenly too small for strolling and Washington Park insufficient for more than a few. However, they play it safe by deciding to place the park far up-island, away from any building likely to occur even in the distant future. A rocky and swampy rectangle of about 750 acres is appropriated by an act of legislature in 1852, although the work will not be started until 1857. They appoint Olmsted superintendent. The commissioners offer premiums for designs. Downing's young assistant, Calvert Vaux, approaches Olmsted, American landscape enthusiast and social reformer, and together they submit a plan under the name of "Greensward." Among many strong contenders, they win.

We know, now, that they were in for trouble. Perhaps the story should stop here and not carry them on through struggles against bureaucrats and politicians. In the midst of quarrels, Olmsted went off to help with the Civil War, as secretary to the United States Sanitary Commission. He traveled west in this position—to Cleveland, Cincinnati, St. Louis, and Chicago—always noticing the "obvious want of a

A view in Central Park, showing the original condition of the land and
the proposed effect. From Henry Winthrop Sargent's *Supplement to the
Sixth Edition of "Landscape Gardening" by A.J. Downing.*

pleasure ground." On personal business in California, he laid out a
cemetery in Oakland and a "village and grounds" for the "College of
California." He was also appointed commissioner of "Yosemite and
Mariposa Big Tree Grove," a state reservation and one of the first seeds
of national conservation. By then the Central Park situation moved
Vaux to beg for Olmsted's return. And return he did, after briefly ad-
vising on a park for San Francisco.

From then on Olmsted's future became a long sequence of parks and
pathways, from the acclaimed perfection of Prospect Park in Brooklyn
to such ambitious and successful schemes as the "Emerald Necklace"
of Boston parks, starting in the lowlands of the Back Bay fens and
linking throughout the city such spots as the Arnold Arboretum,
Jamaica Pond, and Franklin Park. Olmsted's chroniclers today credit
him with thirty-seven urban parks, sixteen community subdivisions,
and fourteen campuses, in addition to his private commissions.

Henry Winthrop Sargent's *Supplement to the Sixth Edition of "Landscape Gardening" by A.J. Downing,* from 1859, gives us a contemporary account of the beginning of Central Park.[3] Sargent began his reference to Central Park by calling it "the most important work of its
kind that has been undertaken in America." He claimed that the edi-

torial articles from *The Horticulturist* urging its necessity and setting forth its advantages "unquestionably exercised an important influence in favor of the project."

The ground set aside for the project was appropriated the year following Downing's death, and its actual purchase was completed in 1857. In early June 1858, the plan submitted by Vaux and Olmsted was adopted by the board, and work commenced. In the most favorable season, as many as 2,000 men worked under the guidance of the designers.

Sargent estimated that the park was two and one half miles long and half a mile wide. Two reservoirs, one old and one new, divided it into two parts. The old reservoir was a quadrangular basin of masonwork. The new was irregular and curved, with earth embankments. The "horizon lines" of the upper park, between the new reservoir and 106th Street, were bold and sweeping, with wide slopes, and needed little alteration of surfaces. A ravine ran through here, in which a small lake might be formed. Where the road met the dip of the two largest hills, a stone bridge would be erected, so the main circuit drive could include a view of all the principal features in the upper park. Sargent predicted that the sixty acres north to 110th Street, consisting of rocky bluffs forming a natural boundary, might be added to the park.

On the easterly side of the upper park would be an American arboretum, "so that every one who wishes to do so may become acquainted with the trees and shrubs that will flourish in the open air in the northern and central parts of our country," not formally arranged but preserving "the most beautiful features of lawn and woodland landscape." As far as practicable, however, the natural order of families would be preserved. In the event that the park might be increased in size, the arboretum would be considerably enlarged.

The lower park, between 59th Street and the new reservoir, more heterogeneous in character, required more varied treatment. A long rocky hillside immediately south of the old reservoir had been accepted as the central point of landscape attraction, to which the other ornamental arrangements were to be subservient. A skating pond or lake of about fifteen acres and of varied outline surrounded a portion of the base of this hill and separated it from the lower park. The summit of the hill was a tableland with expanses of lawn; the side was a

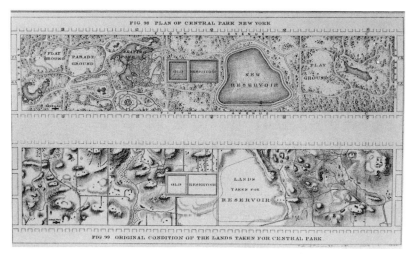

Plan of Central Park, showing original condition of the lands taken.

"ramble" with a labyrinth of footpaths among groves and shrubbery, rivulets, rocks, and glens. The "ramble" included a cavernous passage, which had been excavated and made "an interesting incident heightening the naturally picturesque character of the ramble."

A promenade in the lower park consisted of a broad level walk between double rows of elms with irregularly planted boundaries. Its northern extremity was finished architecturally, elevated to overlook the hillside and the ramble. It connected with the intervening lake by a water terrace, to which it communicated by a flight of stone steps and also through a spacious corridor under one of the entrance drives.

A level parade ground to the west took up about twenty-five acres. To the east of the promenade a "stretch of pleasantly undulating ground" was intended to be planted with fine shade trees. To the southwest of the promenade, fourteen acres of playground were overlooked by picturesque rocks affording ample opportunity for spectators. Different gates at the lower end of the park offered concentration and divergence for those walking, driving, and riding. A short circuit of a mile surrounded the parade and playgrounds. A longer circuit drive passed through the whole length of the upper and lower parks. At the southern boundary, low ground suggested another lake

Fig. 103.—View of Water Terrace in Central Park.

View of water terrace in Central Park and a view of its original condition.

Central Park in Winter. Currier and Ives.

of about six acres. On the western side, a "winter drive" would be protected by evergreens. On the eastern side of the lake, a "geometric flower-garden" of considerable size was proposed.

Because it was possible that the situation of this park might become the heart of the city, transverse roads would be planned. Each would be carried through on a grade that would allow the pleasure drives to pass over the possible future "turbid stream of traffic" by the construction of short tunnels, screened from view by thick plantings.

A "before and after" picture of the water terrace shows the metamorphosis of this "most important American project," inspired by Downing, designed by Vaux and Olmsted, and described for us by Sargent. A Currier and Ives print of New Yorkers enjoying their park celebrates its success.

NOTES TO CHAPTER 12

1 It is interesting to note here that another of our prophets, Charles Eliot, also had a family given to serious touring of their own environment.

2 This rose still flourishes whereas the full name of its introducer has become unnecessary. "Harison's Yellow" suffices. As does "Central Park."

3 Sargent writes from his home, Wodenethe, Fishkill Landing, Dutchess County, New York, an estate briefly described by Downing in his *Landscape Gardening:* "Wodenethe, near Fishkill Landing, is the seat of H. W. Sargent, Esq. and is a bijou full of interest for the lover of rural beauty; abounding in rare trees, shrubs, and plants, as well as vases, and objects of rural embellishment of all kinds." Mr. Sargent generously described the making of Wodenethe in his *Supplement.* The bulk of the *Supplement,* and its main purpose, is a carefully listed description of "The Newer Deciduous Ornamental Trees and Shrubs." We can sense Sargent's personal interest in introducing the most recently tested and proved plantings in the great new park. He regrets having to go to print before some have been entirely proven hardy.

The Final Link: Trees as Keys to

Understanding the Value of Natural Resources

T HE nineteenth century in American gardens saw the origins of many trends, the completion of none, and the abandonment—if one is willing to see it as that—only of "bedding-out," a practice that will be discussed in the next section. One of the seeds planted, later to become vitally important, lay in the idea of conservation. It started, like the first attention paid by the Old World to the New, with trees.

The trees of the American Northern Hemisphere had become important to the countries of the Old World, which had despoiled their own forests. European mountain slopes had been denuded for housing, for warmth, for the building of ships. England had put its great oaks into ship after ship, using straight trunks for beams and crooked or curving limbs for "knees," stronger than carpentered joints. These crook timbers were also used for roofing great halls and barns, so that each high roof, looked up into from inside, seemed like the hull of a great ship, looked down into from the decks. In the seventeenth century, England and other European countries had to travel into Norway and Russia for timbers large enough for ship masts, particularly, which had to be stepped in sections to be tall enough.

And then suddenly, in America, there were stories of trees so tall and sound and straight that a single tree would make a whole mast. King's foresters, sent over to stamp a broad arrow on every tree of suitable dimensions, reported this to be an endless task, because all the trees were at least that size. In New England, huge tree trunks were trimmed in the hills, slid down snowy slopes to rivers, and floated to shores for transshipping.

Many new uses were discovered for American trees. Some had bark suitable for medicine. Many had sap for turpentine and tar, some even for making sugar. Several had beautiful blooms. Others had exotic grains that, when properly cut, yielded wood marked with tiger stripes and peacock-tail eyes. And, of course, countless trees had to come down to make space for people. In the seventeenth century, John Evelyn tried to replace the forests England had sacrificed. In the eighteenth and early nineteenth centuries, officers of the French kings sent eminent botanists to the American East Coast to find suitable trees to replenish French woodlands. We know, from journeys like Manasseh Cutler's, taken by horse and buggy through Ohio forests, that the trees there were so tall and so close that there was no undergrowth to impede sight or progress.

Clearing away forests had become an art—even a mania with some individuals as the population moved west. Mrs. Basil Hall, "in way of seeing sights," was gratified to be able to time tree felling in a settlement where houses had been built among stumps and trees had been girdled to kill the foliage and let sun in upon planted crops. The Halls came upon two men chopping down trees who "good naturedly cut down two for our amusement . . . with a dexterity far beyond anything of the kind in England. The two cut down one tree between fifty and sixty feet high and four feet in circumference in three minutes, all but ten seconds." More sensitive comments upon Canadian forests come from the diaries of Mrs. Simcoe, wife of the first lieutenant governor of Upper Canada at the end of the eighteenth century. Her description of the diversity of American trees comes just after her exposition of a "choke-rope." Travel in winter in New York and Canada was by horse and sleigh on snow-covered, frozen rivers. In thaws, a horse could fall into an unseen hole. A "choke rope"—"learned from the people of the states"—was carried in the sleigh to slip around the horse's neck as he struggled in the water. It was tightened until the horse swelled up, floated, and could be pulled out. And, so, on with the journey—this time a twenty-mile drive through the woods to Kingston:

> I was amused by observing the various barks of trees—the most deeply indented and light-coloured White Ash, the rugged shag-bark Hickory, the regular marked Ironwood, the perpendicular ribbed Cedar, the Basswood, the varieties of Black and White Oak, the Maple, Chestnut, etc., the strong lines of the Pine, particularly the Norway, which is of a rich yellow brown and when cut approaches to a bright orange colour, among all this, the smooth bark of the Beech looked as naked as a frog and had a very mean appearance.

Interestingly, "Norway" shows only what she had seen in England, which a native Canadian pine resembled.

So when word came of trees on the West Coast, larger than any seen before, botanists were sent to collect them and lumbermen moved in to cut them down. Under the circumstances, the early warning against despoiling the continent's resources sounds like a trumpet blast—though one blown in a wilderness.

Fittingly enough, considering where the New World deforestation had started, the impulse to save and replace natural riches began early in the nineteenth century in Woodstock, Vermont. George Perkins Marsh was born in 1801 in a house that commanded a valley, a river, and half the circumference of the sky. It still stands. Wooded hills rise behind the house and beyond the river. It is not a place where anyone could ignore nature or be indifferent to the fate of a nation. His grandfather and his father had been distinguished participants in the new governments of the state and the nation. Young George Marsh had an eye affliction that left him in Vermont, free to wander the countryside for his health. Years later, in 1871, he wrote to Charles Eliot Norton, "The bubbling brook, the trees, the flowers, the wild animals were to me persons, not things." And Marsh added that one would be hard put to "make out as good a claim to personality as a good oak can establish."

Like other young men given time to wander, Marsh tried everything: farming, sheep raising, the law, scholarship, and finally politics. He went to sit in Congress in Washington, where, in addition to his

"The East and the West: The Orient and the Occident shaking hands after driving the last spike," a climax in transcontinental travel.

political duties, he helped to found the Smithsonian Institution. Carrying on politics in Vermont, he helped Zachary Taylor win in 1848 and was rewarded by being appointed minister to Turkey. There, besides pursuing his diplomatic business, he collected for the Smithsonian and wrote a small book suggesting that the camel might be useful in the arid lands of the American West. He came home because of impending bankruptcy in his wool business and an unfortunate railway investment and took to scholarship in a variety of fields, urging a new presentation of history for a democracy—a study of the people. American historians, he said, must treat with "the fortunes of the mass, their opinions, their characters, their leading impulses, their ruling hopes and fears, their art and industry and commerce; we must see them at their daily occupations in the field, the workshop and the market."

Marsh's reputation as a scholar brought him an appointment as minister to the newly formed kingdom of Italy, under President Lincoln. He served in that post with distinction from 1861 until his death in 1882.

Studying the depredations worked upon the European landscape since ancient times, Marsh collected some of his own earlier articles and studies—on the decline of Vermont fisheries and on the destruction of Vermont forests in favor of farming—and combined them with writings on the current worldwide devastation of the environment. His book, *Man and Nature,* published in 1870, is now recognized as a great work, although it came ahead of its time, at a point when everything man could need seemed in plentiful supply. Only far-seeing farmers like George Washington had worried about one-crop systems ruining the soil. A hundred years would pass before the dustbowl would demonstrate Virgil's ancient wisdom in advocating contour plowing. The connection between bared hillsides and soil erosion, or stripped riverbanks and the water supply, would not be noticed by most until hardship made them clear.

Sparked by Olmsted's journeys to California, by the end of the century land was being set aside for preservation in the form of national parks. Irreplaceable wilderness areas of incredible majesty were saved for posterity, like enormous public gardens. Lands over which early explorers and botanists had struggled were set aside for everyone's enjoyment. Greater accomplishments were to take place in the twentieth century, spurred by fears of damage to already protected wild lands threatened by commercial needs. The nineteenth century held only the first awareness of a need to conserve natural beauty, to surround cities with green belts, to build highways through wilderness without destroying it, even to arrange for the average citizen to walk through once fetid swamps.

Another sensitive individual who was able to effect several great innovations and the final link in our chain—was Charles Eliot. Born in Cambridge in 1859, he accomplished a great deal before his tragically early death in 1897.[1] As a young man just graduated from Harvard in 1882, Eliot decided that the "occupation of the landscape architect" seemed to fall in with his "natural tastes and desires," and he entered the nearby Bussey Institute, then the only source of such instruction, to study agriculture and horticulture.

His father and his distinguished uncle, Professor Norton, fellow members of the Saturday Club with Olmsted, then the nationally recognized authority on the uses of public open spaces, interested Olmsted in Charles's future. Charles gladly accepted an apprenticeship in Olm-

The Natural Bridge ("Perspective taken from Point A"). From *Travels in North America in the years 1780, 1781, and 1782,* volume 2, by the Marquis de Chastellux, translated from the French by An English Gentleman who resided in America at that period. (London: G. G. J. and J. Robinson, 1787). Charles Eliot's first paid job (for his keep).

sted's office and was able to "ride out" with Olmsted, all over Massachusetts and on to New Jersey, Washington, and Detroit.[2]

Olmsted's office at this time was developing Franklin Park and the Arnold Arboretum, two components in the plan for an Emerald Necklace linking a series of beauty spots across Boston. Charmingly coincidental for us, considering the deference paid to the location by Jefferson at the beginning of the century, was Eliot's task of spending a week in 1885 at the Natural Bridge in Virginia. Here 2,000 acres were owned by a Colonel Parsons, who hired his young advisor for room and board and provided Eliot's first paid job. In return, Eliot spent his time, he wrote his mother, attempting "only easy work giving immediate effects—breaking up straight edges of woods, opening vistas, clearing to bring out fine trees and opening lines through the woods for new roads."

Arthur Shurcliff's sketches: Before and After, "Tree-clogged notch, near
the southeastern escarpment of the fells, which might command the
Malden—Melrose valley and the Saugus hills."

Olmsted had become closely attached to the leading Boston architect, H. H. Richardson, who built in what seems to some of us today a massive scrapbook-romanesque style, which even in its first unveilings required careful attention to its surroundings. Using assorted motifs from a past we feel we have inherited demands complete subservience in the surrounding landscape, as those of us who have tried to show Richardson at his best understand—horticultural confusion is to be avoided at all costs. The young Charles Eliot may have been a tempering factor in potential gales, leaning as he did toward noncluttering simplicity. The idea that simplicity was desirable came as a daring innovation in the nineteenth century. Eliot's vision of direct involvement in the basic beauties of any site is apparent today in the sequence of parks and drives around Boston. His dedication to revealing the existing landscape to its greatest advantage is clear from our before-and-after sketches, by Arthur A. Shurcliff, of the proposed development of the Fells in the Malden–Melrose valley.

By the end of 1885, young Eliot went abroad, to study as hard as he would have had to in attending the nonexistent, much-needed school of landscape architecture. His letters home were so entertaining and enlightening that Olmsted insisted Eliot include writing for the general public as part of his professional calling. The Eliot family, like the Olmsteds, had been used to taking family excursions, and Charles was able to equate for them the features of the coasts of Europe with those of Maine, and Denmark's shores with those of Massachusetts.

After his return to the states, Eliot conceived of and saw to fruition a plan for a privately supported organization for preserving "beautiful and historical sites" throughout the Commonwealth of Massachusetts. In urging its acceptance by an enlightened group of Bostonians—similar to those responsible earlier in the century for a landscaped cemetery and a horticultural society—he began:

> There is no need of argument to prove that opportunities for beholding the beauty of nature are of great importance to the health and happiness of crowded populations. As respects large masses of the population of Massachusetts those opportunities are rapidly vanishing. Many remarkable natural scenes near Boston have been despoiled of their beauty during the last few years. . . . Scattered throughout the State are other places made interesting and

valuable by historical or literary associations; and many of these
are in danger.

The resulting Trustees of Public Reservations, incorporated in 1891, is
the oldest conservation agency in the United States and is still a growing power. It was copied, not only all across America but in England
for the creation of their great National Trust.

Eliot also steered through to success the creation of a Metropolitan
Park Commission, in 1893, originally, as Eliot reported in his description of the "General Plan of the Metropolitan Reservations" in 1895,
"in order that some of the more striking scenery of the district surrounding Boston might be preserved for the enjoyment of existing and
coming generations."[3]

IN studying the origins of the Arnold Arboretum, near Boston, we are
considering a living monument to the accomplishments of the nineteenth century, a prism that brings together the lights from all the
seekers we have followed across the miles and through the years. It is
one of the phenomena in the development of the new democracy of
the United States that the pattern of civic responsibilities for amenities
as well as necessities moved steadily westward. The sequence of events
that resulted in the Arnold Arboretum was repeated and moved westward as surely as the all-purpose English cottage gardens of the seventeenth century had made their way in ox carts westward with frontier
families.

In 1842 Benjamin Bussey willed to Harvard College 394 acres of
varied countryside in West Roxbury, near Boston, to be used for a
"school of agriculture and horticulture." The land was left subject to a
life interest, so not until 1869 was a school building erected on seven
acres released for that purpose. In 1871 the school opened with a faculty of five, among whom, for the first year, was the leading authority
on rose growing and eminent historian, Francis Parkman.[4]

At about the same time, the aging Asa Gray, teaching natural history
in another department of Harvard, longed for an arboretum, prefera-
bly near the Botanic Garden so ably served years before by Thomas
Nuttall. Gray was working with Torrey to finish his *Flora of North
America* and was reluctant to move far from his desk and his laboratory.
Fate intervened with another bequest to Harvard.

James Arnold (1781–1868) of New Bedford, a proficient gardener who had been visited and praised by Downing, left a quarter of his estate to be administered by three friends for "the promotion of Agriculture and Horticultural improvements or other Philosophical or Philanthropic purposes at their discretion." Fortunately, the three friends included George Barrell Emerson (who had written the great and popular *Report on the Trees and Shrubs Growing Naturally in the Forests of Massachusetts* in 1846) and John James Dixwell, who took pleasure in raising trees on his nearby estate. The third friend, Francis E. Parker, was an able Boston lawyer.

Although Gray wished the arboretum to be in Cambridge, land had so increased in value (in the phenomenal rise that was to wipe out the whole Botanic Garden in the next century) that the bequest (about $100,000) would only have acquired the necessary land. Instead, an arrangement was made to turn over part of the Bussey land for the establishment of an arboretum, using Arnold funds.

In 1872 Charles Sprague Sargent, who had learned all he could about horticulture from Asa Gray, became curator of the Arnold Arboretum. By 1879 he was settled into his lifelong career as director of the Arnold Arboretum and Arnold Professor of Arboriculture. While Frederick Law Olmsted created his great park system for Boston, Sargent improved the Bussey land under Olmsted's tutelage. Astutely, he joined the arboretum to Olmsted's series of parks, in return for municipal help with maintenance.[5]

The arboretum today continues its attention to the general public and its joys in the seasonal features of its grounds, still obviously touched by the hands of the early planners. Its reason for being, "to collect, grow and display as far as practicable, all the trees, shrubs and herbaceous plants, either indigenous or exotic, which can be raised in the open air at the said West Roxbury," ties it forever to our gardens. So our trees continue to flourish with their *expressions* and *personalities,* as savers of our environment and tributes to the great gardeners and planners of the American nineteenth century. There is no possibility of doing justice to these great men, running before their time, and even now, in many cases, still before ours.

NOTES TO CHAPTER 13

1 His father, president of Harvard College at the time Charles Eliot died, commemorated him in a warm book, published in 1903.

2 Many years later, Eliot gave Olmsted credit for having coined the term *landscape architect*.

3 Planning for future betterment was so strong that when, shortly after Charles Eliot's death in 1897, his father went to call on another son, Samuel, whose wife had just given birth, he asked, "A boy or a girl? A boy. Good. His name will be Charles and he will be a landscape architect." So it was.

4 Within a year, he was succeeded by Charles Sprague Sargent, also a neighbor to the property and a nephew of Henry Winthrop Sargent, who eulogized Downing and explained to us how to make country places. Charles Sprague Sargent became identified with the final disposition of the area and wrote the great *Manual of North American Trees,* published in 1905.

5 As we watch the gathering of personal forces here, it cannot overly surprise us to find Mr. Jack of Montreal, whose wife's advice we have studied, also working at the arboretum. In 1884, when the arboretum's plans, originally made by Olmsted, had to be enlarged and revised, Charles Eliot, still an apprentice, made the new drawings. As an added touch to help tie our threads together, the original Arnold house in New Bedford was taken over and made into a "French château" by William J. Rotch, whose own house, a "Cottage Villa in the Rural-Gothic Style," had been designed by Downing's architect friend, Davis, known to Vaux and described by Downing in his *Architecture of Country Houses*. Ironically, as the great botanist Dr. W. T. Stearn pointed out when he came from London to speak at the hundredth anniversary of the arboretum's founding, a wry justice appears from the fact that the Arnold fortune had come from lumbering Michigan forests.

Domestic Pleasures

FOURTEEN

Tracing the Outlines of the
Past: Living Clues and Eyewitness Accounts

T is now time for us to come out into the gardens themselves. Living clues are always superior to inspired guesses when we try to come as near as possible to the gardens of the past. Plants have their own powers of enrichment when they can divulge them. A plant that was given to me as a child as being such a staunch friend it would show no signs of having been transplanted (blue Greek Valerian, now *Polemonium reptans* Linn.) holds a special charm because it came from a massed planting in Portland, Maine, which had been resorted to during the War of 1812 to hide where the family silver was buried when it was feared the British were coming into the harbor.

Contemporary garden plans that were drawn up by owners and dated have a special heartwarming value, as witness Thomas Jefferson's repeated sketches of his desired roundabout walk at the back of the house at Monticello, planted on each side with a border of special flowers (a plan well worth carrying out today, although there is said to be no archaeological proof the design was ever actually followed). A garden design with its original borders of stone or brick still in place and easily recognizable is a gem today. We are lucky to be able to present such plans here and to add several others with a veracity that, in these days of what amounts to overrestoration, is hard to come by.

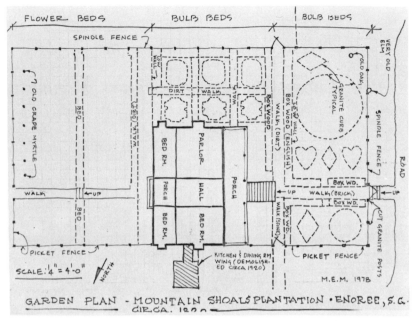

FLOWER BEDS   BULB BEDS   BULB BEDS

Plan of Mountain Shoals Plantation. Courtesy Frank Coleman.

AT Enoree, Spartanburg County, in up-country South Carolina, is a plantation house built in 1818–23 by James Nesbitt, descendant of the first twenty Scots-Irish settlers who came to the area in 1751. Enoree is an Indian name for the muscadine grape (*Vitis rotundifolia*), one of the wild grapes of the South that attracted notice from the time of Sir Walter Raleigh's colony.[1] The plantation, which was to grow to 1,400 acres (and, as family tradition states, 200 slaves), was named Mountain Shoals for the shallow ford in the nearby Enoree River. Constantly lived in since its construction, the house is today occupied by a descendant of the original builder who tends the small garden, laid out in a design elaborate even for an enthusiastic gardener today. The plant material has survived to furnish a wealth of witnesses to what was expected of a garden in the early nineteenth century.

The design has charm. Surrounded by a variety of fencing and divided by a number of walks—typical of the time and quality of the garden plan—the garden beds are also varied in shape. Carefully marked out by small granite slabs are beds shaped like diamonds, hearts, and ovals and a curious design of a rectangle with six semicircular indentations, one at each corner and one in the middle of each long side.

The fencing in the formal front has granite posts and granite stringers to support a picket fence with a small diamond-shaped point to each picket. On one side, the pickets change to spindles (recommended in books of the times as sturdier than pickets). The main walk, from the front gate, is brick with box borders. At the front steps, walks and box borders turn to each side to run across the face of the house. The walk on the shorter side, toward the former kitchen, is of stone. The walk toward the garden side of the house is of dirt. Both are box-bordered. The walks in the small, elaborately laid-out garden by the parlor side are of dirt. This is obviously a garden designed to be enjoyed.

At Mountain Shoals Plantation today, two interesting plants attest to the tastes of the "curious" former owners of the garden. One is the "feathered hyacinth," which grows in profusion in the granite-bordered beds. *Muscari comosum monstrosum,* as it is called today, arrived in England from southern Europe in 1596. It was offered in M'Mahon's nursery list in Philadelphia in 1806 and sent by M'Mahon to Jefferson in 1812 as "Feathered Hyacinth roots, *Hyacinthus monstrous L.*" The other is the "green rose," *Rosa chinensis viridiflora,* brought from China years before it was marketed in 1856, a curiosity with a bloom like a bursting bud, not pretty but interesting to have. This is said to have been its first appearance in this country.

Here it may be well to say a word about gardens generally, and American gardens in particular. With so many different conditions all over the country, it is folly to try to generalize; the interest lies in the peculiarities of each area. When Washington rode around the newly created country, visiting the thirteen often very different states and taking note of the soil and growing conditions in each one, his comments about the promise of each state were succinct to a degree—and helpful to us now, as we research the gardens. When he commented upon the superior agricultural practices of the New England states, he added rather wryly that the success must be credited chiefly to the industry of the inhabitants, because the whole countryside was under

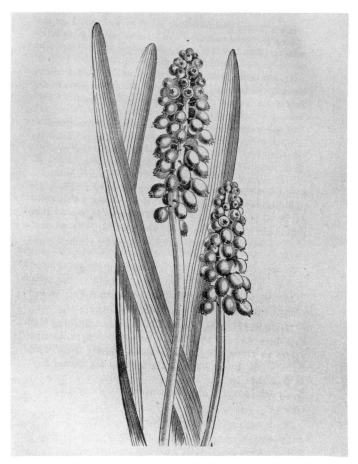

Grape hyacinth (*Muscari moschatum*). A close relative, the larger
"feathered hyacinth" (*M. comosum monstrosum*) is a feature of
southern gardens. From Breck's *New Book of Flowers*.

snow for nearly five months every year. We take this as a fair directive
for judging gardens before the current demands made upon them by
tourist traffic.

For there is no doubt that tourists like total effects—color and mass—
which explains the prolonged existence of "bedding-out" in public
parks and the gorgeous layouts in front of summer hotels. It also explains the supposed duty of historic shrines to lay on mass effects to

please the weary tourists descending from crowded buses. Acres of tulips *are* refreshing. Hyacinths by the hundreds cannot be forgotten. Chrysanthemums by the yard are cheering at the end of the summer. But they are all tourist-oriented. The dedicated gardener who seeks an unrestored but still preserved historic garden of any date but the present is out of luck, except in places like Mountain Shoals.

Here the garden and its plants show what things were like before gardens had to be in top-notch bloom all the time, regardless of climate. Here one can perceive that the spring was a lovely, long, slow affair, full of fragrances and bright with bulbs, early blooming shrubs, and violets. Summer brought roses and the scent of box. The end of summer showed the late lilies and members of the amaryllis family, and the little fruits of the gay red hips of roses and the shining white berries of the snowberry bush. And that was that. With ground-covers like periwinkle, borders of thrift and pinks, and edgings of box, the plan of the garden is secure throughout the year. Its effects are to be awaited annually—like early strawberries and asparagus before canning and deep-freezing enabled us to eat things out of season.

The garden at Mountain Shoals is a valued pleasure garden, with the box increasing to form hedges to walk between or to see over, as far as the river in the valley below and on to the far hills. That the family intended the plots for their own enjoyment and not to please the passing stranger is proved by the original fencing, where the sleepers are on the outside and the pretty palings face into the garden—a lesson for us all.

There are a few obvious rules for gardens like this. Symmetry is paramount. Crape myrtles at the sides balance each other. Each bed of sufficient size is centered with an ornamental shrub. Roses mass behind the box hedges that lead to the front door. The scale is modest and consistent. Several garden novelties are entertained—for instance, the "green rose."

As all good gardens are loved and remembered later and elsewhere, we are fortunate to have a record of a garden made by a daughter of the house at Mountain Shoals who grew up there, married, and went forty miles away to a garden of her own, which was apparently as like as possible to the garden she had grown up in. *Her* daughter has left us her written memories of that garden, made in 1868, gone but recol-

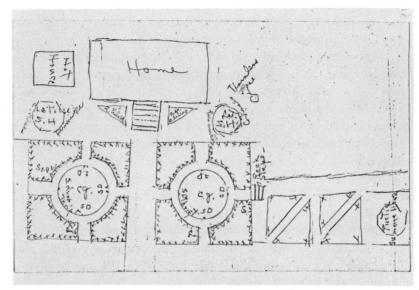

Plan of the garden at Williamston, created by a daughter who grew up at
Mountain Shoals. Drawn by *her* daughter, Andrena Anderson Parker.
Courtesy Frank Coleman.

lected with freshness and affection in 1956. Remembering the house at
Mountain Shoals, the granddaughter of James Nesbitt writes of her
mother's garden in Williamston.

I feel sure that there was an especial appeal to the house for
Mamma . . . white, on a hillside and big boxwood bordering the
path to the door as in her girlhood house . . . at Enoree. She had
made a garden of the formal English type in front, a circular bed
on each side was surrounded by four beds each bordered with
box. . . . Only there was thrift around the circular beds—in the
center of each was a large gardenia (Cape Jessamine, we said then).
There were clusters of snow-drops beyond the pink thrift. She had
bulbs, all kinds in various places, especially in the lower garden
where the beds were triangular—lovely daffodils, narcissi, and
hyacinths. . . . She loved most the snow-drops and the pale blue
Roman hyacinth—single. I have drawn very poorly the plan so
you may have a clearer idea. Of course there were roses, dahlias,

lilies, golden bells, French honeysuckle, syringa, snow-balls etc. thoughtfully and rather symmetrically arranged—at least with a sense of balance. Where she had lilac crepe murtle on one side you would find some to balance on the other. When I think of the size of that box leading to our door-steps it must have been five feet tall. I think of it as taller than I. The flower beds had rock sides—the soil filled in. The walks were white always with sand brought up from the creek. Being accustomed to it always, I did not notice the fragrance of the box wood, but I remember when a child, hearing visitors say it was delightful. Of course the x's indicate boxwood. The summer house of lattice was gone years ago. . . .

I don't recall seeing elsewhere the golden yellow thornless rose. I think its name was Rose of Sharon. The blooms were small, grew up the stem—like golden bells. . . . The favorite periwinkle, dark green, slender waxlike leaf. . . . August lilies at the foot of the steps. In the garden there were lemon (lilies), Madonnas and a Japanese lily that was much admired . . . except in coloring, similar to a tiger lily—the petals were white to soft pink, dotted with dark red. . . . Of course there were violets—both purple and gray. . . . Largely tea and pink roses, but there were the Cherokees and the Seven Sisters also.

A postscript says the box was sold to the Williamsburg restoration for a reputed $1,500 and that the writer was heartbroken until she saw the garden again, deserted and neglected, and realized her mother would have been happy that the box "is now cared for and enjoyed."

Because the memories of contemporary observers are always worth recording, we insert here a bit of garden design not generally found in books. We know that George Washington concerned himself with building four garden structures like traditional pepperpots, two to stand near the house in the garden layout and two to go at the far end. One at the far end was designated a schoolhouse for tutors and children; the other was probably used for tools and storing bulbs. The two near the house were "necessaries," or outhouses. The point to be made is that garden structures had many uses. There were usually a wellhouse, a summerhouse, and a tool and storage house, which could double as a summerhouse when it stood high, like the one remaining structure in Lady Skipwith's garden, from the eighteenth century.

There was also sometimes a bathhouse, and there was always an outhouse. But here we find a recorded difference in uses. At least we have been told, by a lady brought up on a southern plantation, that the outhouse at the end of the garden was for the ladies of the household. Men were sent the other way, to conveniences near the stables. This plantation-reared lady remembered that when she was a child the plumber came to live with the family for several weeks to accomplish the taxing job of installing modern facilities. And then, even after he installed a real tub in a former bedroom that had been set aside for the bathroom proper, her mother continued to bathe as always, in a large saucer tub on her bedroom floor, by a fire that heated large pots of water and with kindly black hands to pour the bath. And she continued to walk through the roses in the garden to reach the outhouse, where the path was planted with bulbs to make her passage more pleasant. There is no record that the men's accommodations were similarly landscaped.

This explains the design of many southern gardens, where the weather made outdoor facilities possible the year around. In the North, as we know from Cotton Mather's regrets about time lost merely sitting in the orchard, the men at least went to a shelter there. Later it became customary to connect a succession of sheds to the kitchen end of snow-conscious homesteads to afford shelter all the way to the outhouse for all.[2]

For landscaping the outhouse was a factor in garden design. When the propagator of the Concord grape, Mr. Ephraim Bull of Concord, Massachusetts, discovered a "chance seedling from a chance seedling" that bore such promising fruit he exhibited it in 1853 at a show put on by the Massachusetts Horticultural Society, he began to entertain grape-minded visitors. His great-granddaughter told me that when a new grapevine appeared to grace the already vine-covered outhouse, no one paid it much attention until it bore very fine red grapes—by then too late to credit the donor.

ANOTHER southern garden for which we have a plan and documented evidence of the planting is the one Jackson had made for his wife in 1819 at the Hermitage in Nashville, Tennessee.

It is a large garden of more than an acre, which General Jackson had placed close to the house to be enjoyed by Rachel from her bedroom

Plan of the Jackson garden at the Hermitage, showing outhouse and tomb. Drawn by Pattie Hall from picture provided
by the Hermitage Board. From the Hermitage brochure.

windows. Laid out in four large squares around a centerpiece of five circles, separated by four shaped triangles, one of the walks leads back to a "necessary house" set at the end between two large squares.

The original garden was designed "by an English gardener named William Frost" and was laid out to be visited by the many callers who came to pay their respects to the general and his wife, who was known to love flowers. In her short ten years in the garden she was remembered as always proffering a nosegay to a departing guest.

When Rachel died in 1828, her husband had her buried in one corner of the garden and had the square there cut up into three sections, the one in the corner accommodating a handsome tomb over Rachel's grave and later his own, with surrounding trees and shrubs.

Several attractive accounts show us the garden as it was, both when

Rachel lived there and again after she had died, when the venerable Jackson, retired from the presidency, lived there again alone. One guest who arrived in 1827 with others, "in the barouche, servant mounted," wrote that they "rode up a long avenue" and were presented to General and Mrs. Jackson at the hall door. Mrs. Jackson led them into the drawing room, insisted on their taking some "refreshments which were handed," and, after they were rested, "proposed walking into the garden, which is very large and quite her hobby. I never saw anyone more enthusiastically fond of flowers. She culled for me the only rose which was in bloom and made up a pretty nosegay."

Another lady, born in 1840, who remembered the garden from the time of her childhood, could recall "some of the flowers, shrubs, etc. which were growing when General Jackson lived there" and obligingly made a list that took special note of the planting around the tomb— magnolias, weeping willows, and crape myrtle. She was the daughter of Colonel Parker, who operated the hotel that was named after General Jackson and rented from him. The two men were friends, and she was often in the garden.

Some of the flowers she remembered as growing when General Jackson lived there were "old-fashioned pinks of all colors"; peonies—red, pink, and white; verbenas; poppies; sunflowers; hyacinths; and tulips. Tiger lilies grew beside the walks, and there were white lilies also. The shrubs were "several big crepe myrtles near the tomb"; "a Japonica bush" (*Chaenomeles japonica*); a calycanth (*Calycanthus floridus,* or Carolina allspice); white and purple lilacs; "old-fashioned honeysuckles, coral, white and yellow." There were many roses—cinnamon, tea, moss, and "wild." Under both the shrubs and roses she has listed "Mycrophellia," a sign of its importance. This is interesting because the rose called *R. Microphylla* arrived in England from China in 1828.[3]

Those who like to find poignancy in some plantings will find one in General Jackson's roses. "The Cherokee Rose," arrived from China, became so well acclimatized it was considered wild all across the southern territory and was given the name of the Indians Jackson pushed before him into the hinterland.

Another visit was recorded in a diary of 1851, by the wife of Captain Dufield, owner of the steamboat "Talleyrand," who had been invited to visit the Hermitage by the then-owner, Dr. Jones. Jackson had

The everblooming Chinese rose, one of the Far Eastern
introductions, which brought a new wealth of roses
to gardens where they could withstand the climate. (Also
called the "Bengal rose.") From Curtis, *Botanical
Magazine*.

died in 1845 and been buried beside his wife, with only his birth and
death dates on his slab set into the square column in the center of the
circular pillared structure. The inscription on Rachel's slab describes
her many virtues and is obviously written by an adoring husband.

Mrs. Dufield writes that the

garden in which the ornament is erected is beautifully laid out
with flowers and fruits. There is a small circle in the middle of
which is one solid bed of verbenas, pinks, tulips, pinys and other

Old-fashioned pink, as in Jackson's garden. From Curtis, *Botanical Magazine*.

flowers too tedious to mention and too beautiful for me to attempt a description.

The house, she says, is "situated some distance back from the road,"
with a "circular" in front and smaller circulars on either side, all three
laid out in "flowers and cedars." The balance of the yard in front is
filled with cedars and forest trees, under a "cluster" of which the party
sat down to have its lunch.

And one final quote from a visitor, this time a granddaughter named
Rachel Jackson Lawrence.

As you enter the garden gate you find the fringe tree planted by Grandpa. Passing down the walk you find the crepe myrtle on either side. Along the border farther down are lilac and syringa. At the far side of the middle plot there is a smoke tree, and following the path down you find lilac, crepe myrtle, mock orange and along the back fence of the garden these same shrubs.

As you enter the garden on the left hand side is the calicanthus or Sweet Betsy. There are the fig bushes, the flowering almond and many other shrubs. Around the tomb are the magnolias.

In the garden the flowers were the June lily, lily of the valley, single white and blue hyacinth, the red, the white and the pink peonies, blooming in succession as named.

The center beds are filled with the old-fashioned sweet-williams, petunias, periwinkles, blue-bells, pinks and other garden flowers. There were iris and jonquils and, as we called them, golden candle-sticks. There was the coral honeysuckle which hung in great clusters on the right side of the formal plot. Among the roses there was the old-fashioned little yellow rose, the hundred leaf pink rose, the moss rose and the large white cabbage rose. There was the Japanese magnolia planted near the center, the rare cucumber magnolia, the very rare tree peony. Like all old gardens in the corners were violets and blue bottles. Box trees marked the corners of the walks. There were several evergeens around the tomb and a bunch of hickory trees planted by Grandpa. Now this is all I can remember.

In 1852 the keepers of the garden were "bricking around the beds" and rejoicing in a supply of fine roses, "about fifty varieties."

We have almost seen this garden grow in the thirty years of recorded observations. Time has been stretched—earlier to include Chinese imports, later to see the arrival of petunias and the Harison's yellow rose. One of the constants has been the so-called Roman hyacinth. The "golden candle-sticks" may be the *Asphodelus luteus* of Miller, listed by M'Mahon, grown by Lady Skipwith.

THE last of the four gardens of the first half of the nineteenth century for which we are able to present original designs and contemporary accounts of plantings is at the Longfellow House in Cambridge, Massachusetts. But before we consider the American poet's plans and plants, let us ponder the general similarity of these relatively far apart gardens—all "formal," two even designated as "English," all dependent upon a mixture of shrubs and flowers in neatly geometrical, edged beds. These are like the gardens of the eighteenth century: Lady Skipwith's plan at Prestwould, walled with "tabby"; William Byrd's brick-walled garden at Westover; and inside the brick walls of the pointed oblongs made by George Washington at Mount Vernon. These four more recent gardens are not all walled—some are only lightly fenced—but these fairly modest "pleasure plats" have an architectural solidity. And the plant materials are similar—all fragrant, tinted with soft colors, and set out to stay and grow old in place. When we realize what is ahead in garden design—the dependence upon "new" annuals, color contrasts, glass frames, lawns sheared by mowing machines instead of scythes, curves and recurves in edgings, suburban developments—we can appreciate the "time lag" factor in what Americans liked to grow close to their homes.

In 1843 Henry Wadsworth Longfellow, then thirty-six, already famous and married to his second wife, Fanny Appleton Longfellow, took possession of the Craigie House on Brattle Street in Cambridge, Massachusetts. It had been purchased for them by Fanny's father. Longfellow had lived in the house since 1837 as a young professor, in rooms rented out to carefully selected young men by the rather eccentric Mrs. Craigie, who wore a white turban and a gray gown and who could snap her gray eyes with telling effect. He became attached to the house and its grounds overlooking the Charles River. It had been built by a Tory, had served as Washington's headquarters during the Revolution, and was being lovingly preserved by the widow Craigie, who, Longfellow felt, neglected the grounds. Annually afflicted with "rose fever," she apparently never gave up roses and insisted upon suffering even the cankerworms to have their way with the old elms, because, as she explained to the anxious poet while the worms dropped upon her white turban, "Why, Sir, they have as good a right to live as we; they are our fellow worms."

Free at last to do as he pleased with the house and five acres of ground, with another four acres across the street "to keep the view open across the marshes" and to the river, Longfellow set about to improve the property. He tried to preserve the remaining elms and to re-place those lost "by transplanting very large trees," some from Water-town and Lincoln, two from the college grounds. Fanny wrote lightly of the fun they were having "drawing maps of our estate, which we decorated with rustic bridges, summer-houses and groves." There was to be a linden avenue, "which my poet intends to pace in his old age and compose under its shade." To her, she says, he is "resigning . . . all the serpentine walks, where in the abstraction of inspiration, he might endanger his precious head against a tree." They were well aware of curving walks and trees in groves after happy experiences living abroad but were able to forgo copying what they had seen. A "linden avenue" at the back of the property on the northern boundary was hoped to prove useful "in screening us from any unsightly buildings Mr. Wyeth may adorn his grounds with." Fanny puts the case for all New England gardeners by telling her correspondent that "out-of-door work is sealed by the frosts until spring, and consequently within doors it begins to flourish."

The next year "a famous turk's turban of box dignifies the front door-steps," generously bestowed by "a very near neighbor." As "roses are to bloom behind them," we may assume there were two clipped and shaped turbans, one on either side. Work goes on. Vines are set out to screen the eastern piazza. The balustrade along the roof has had to be replaced, and the sound parts have been transferred to become a railing on the piazza. Fanny is pleased with the fact that "our externals are beginning to look a little tidier."

In 1845 Longfellow noted that "through Mr. Gray" (to us, the fa-mous Asa Gray) he had imported from England "a number of ever-greens, among them a cedar of Lebanon and pines from the Himala-yas, Norway, Switzerland and Oregon." These were planted in "quite a little grove of little pines and hemlocks." The gravel walk around was continued, and a hedge of sweetbriar and barberry was set out. He built "the rustic seat in the old apple tree" and set out "the roses" under the library windows.

And now, still in 1845, he "made the flower garden, laying it out in the form of a Lyre." He had made a plan showing the house, a new barn, lilac bushes, fruit trees, and an acacia hedge, "given me by George Owen, my publisher," by a side wall. All of this plan remained for years (the acacia hedge, indeed, growing into a row of trees), but the lyre-shaped flower garden apparently proved unworkable. However, the "green trellis work," which had been part of "an old covered walk to the outhouses," had been "set by the flower garden," where the gateway was from "the old college house which stood opposite the bookseller's in the college yard." Longfellow felt it worth noting that under the floor of the covered walk to the outhouses they had discovered the skull of a dog, with a brass collar marked "Andrew Craigie."

A new garden design was made by one Richard Dolben in 1847, reported as being a circle inside a square in the center and two oblongs on either side. This was further elaborated by cutting the circle into four tear-shaped garden beds and the oblongs into quatrefoils. These are submitted here, with the feeling that whatever the Longfellows finally decided upon was probably modified to agree with their own enjoyment—box-bordered triangular beds in the corners with shrubs, and circles with centers of roses. Fanny made a reference to the garden layout as like a "Persian rug," which would explain why the four little pointed ovals that make up the center design are more like gourds than the conventional bed patterns of pears or teardrops. There may have been a sundial in the center, although the one inscribed with one of Longfellow's favorite quotations (from the Italian of Dante, translated as "Think that this day will never dawn again") was placed there in our century, long after his death in 1882.

And again, we are indebted to near neighbors and visitors who could remember the garden. One of these was a child who visited her grandfather, across the road from the Longfellow house, "next Professor Longfellow's meadow." The house there had been occupied by Washington's medical staff when he was quartered in Craigie House. Her happy memories of the area are filled with definitions of associations—the old willows marking the site of the moat around the palisades built by Governor Dudley against possible Indian attack; Andrew Craigie allowing Mount Auburn Street to be cut through his estate after the Revolution; Washington visiting in the arbor. There was an original sort of summerhouse retreat on her grandfather's place, which antedated the 1860 iron octagonal summerhouse shaded by a pear tree. The earlier summerhouse had been built by her uncle when he was at Harvard in 1843 and was made "of saplings with the bark on

Longfellow's "Persian rug."

planted upright in the ground. One side was the door, the other three had diamond-shaped windows, and it was paved with beach-stones." Near the house was a "retired arbor," where one hoped Washington had sat, made by placing a circular wooden bench around the trunk of a giant cedar and, around that, farther outside, a high hedge of huge lilacs.

Among the garden's charms were "rockwork" of discarded building stones set with a center "cuplike" pot, planted with ferns, surrounded by a ring of six or seven others full of wildflowers brought in with leaf mold and rich earth from the woods. A miniature garden behind the summerhouse was laid out with a walk, around York and Lancaster roses in the center. Another tiny box-bordered garden near the porch held moss roses and mignonette. And an important note was the beauty of the west lawn before the scythers came to cut the grass—full of clover, buttercups and oxeye daisies, blue chicory, and tall grasses in flower. Bulbs were in an oblong bed on one side. Grass paths divided

Formal New England "homestead" garden. Courtesy New Bedford
Whaling Museum.

the asparagus and strawberry beds. The lawn near the house was reserved from planting for games of croquet, tennis, and archery. The
main entrance was through wooden gates hanging from granite posts
in a brick wall, capped with boards set to form a V. The "garden
proper," reached from the entrance to the front door (where the steps
had pots of agapanthus and a "red iron vase of classical shape full of
flowering plants from the conservatory"), was a "Broad Walk" almost
all the way across the property, bordered by shrubs in wide beds: *Pyrus
japonica,* snowberries, spirea, smoke bushes, syringas, rose bushes,
Missouri currant, and snowballs. And among the shrubs were planted
larkspurs, monkshoods, snapdragons, Canterbury bells, foxgloves,
dahlias, hollyhocks, stocks, chrysanthemums, yuccas, salvias, honesty, sweet williams, and lilies (Japan pink and tiger). More lowly
flowers included the old-fashioned pinks, heliotropes, Solomon's seal,
spiderwort, bachelor's buttons, and dicentra "with its strings of heart-
shaped pink blossoms" (which makes it *Dicentra spectabilis,* then only
recently imported from Japan).

With this wealth of planting information next door, we can rest on Longfellow's heralding of the buttercups in the grass, the "horse-chestnuts lighting the landscape with their great taper-like blossoms," walking before breakfast "to inhale the air from the cherry-blossoms and to drink the first foam of spring," and, finally, "lilacs in bloom and the apple trees—the whole country is a flower garden."

NOTES TO CHAPTER 14

1   It was collected and grown by Jefferson at Monticello and appears in an 1821 letter to him from Samuel Maverick of South Carolina. The scuppernong grape, popular for winemaking, is a variety of muscadine.

2   And here we may also mention that though there were often three holes, they were cut to different sizes for a man, woman, and child and not for simultaneous occupation. Enough of modern misapprehensions.

3   It was supposed to be related to the Macartney rose, although it is now regarded as a species. It has been called the "Chestnut rose" (*Rosa roxburghii*) and "Old Purple" (grown by Lady Skipwith, I think). The interest to a rose fancier is that some of the old roses in southern gardens—or even those considered as growing wild—are really early Chinese imports: the Cherokee (*R. laevigata*), Banksia (*R. banksiae alba* and *lutea*), Macartney (*R. bracteata*), and the "green rose" (*R. viridiflora*), which seems to indicate they may have been brought directly from China and not sent from England. Charleston, South Carolina, was an ideal place for the tender "tea" roses from China and actually the site of the first combining of a tea or "everblooming" Chinese rose with a once-a-year-blooming old European rose. Mr. Champneys of Charleston crossed a musk rose (*R. moschata*) from the Middle East and Parson's Pink China (which today still grows in Bermuda). The Noisette brothers, great French nurserymen, took Champneys's Pink Cluster to Paris and started the whole tribe of Noisette roses from this initial triumph.

FIFTEEN

The Uses of Space, and of

Embellishments

SES of space and interpreta-
tions of its importance and function are what most simply and clearly
demonstrate differences in the first three centuries of American gardens, before twentieth-century architects, keen upon enclosing space
as if it were solid, made it into a feature to be celebrated for itself.

In the seventeenth century, space, once cleared, was used to its
utmost capacity. Gardens were compact, designed to be ready to hand,
planted closely with everything anyone could possibly require in a
hurry. That they were also pretty and gay was an accident of fortune.
The attention paid to the arrangement of the plants depended upon
how they were to be harvested—pulled bodily out of the ground or cut
back to produce leaves, petals, and seeds. Raised beds were essential at
first, kept in shape—or, as they said, "in fashion"—with boards or
bricks or stones and then built up with edgings of tough little plants
that could be clipped or that had naturally thick and ground-clinging
leaves. A tub for water and some movable benches to make garden
tasks easier were all the furniture required. The wellhead acquired a
bucket on a rope, a raised and fenced-in rim, and a small roof to cover
it, instead of the earlier well sweep, with a heavy rock secured to one
end of a long sapling balanced upon a forked tree trunk sunk into the
ground. Summerhouses had to wait only upon a time when there

American Homestead, Spring. Currier and Ives. Note the wellhouse, and
see open area under porch—later to be latticed and then hidden by
"foundation planting."

could be control of enough space to allow some of it to be set aside for
nonproductive and recreational activities. All fencing was totally efficient—upright split palings or horizontal split rails or boards placed
close enough to discourage wandering pigs and poultry.

With the eighteenth century and the idea that walking in the garden
(where else was there for ladies to walk?) was good for the health,
walks were widened and paved with brick or gravel. There were seats
to rest upon at intervals and at the ends of flower-bordered main paths.
Wellhouses were covered attractively, often to match summerhouses,
and summerhouses became almost that—places in which one could
spend hours, play games, converse, romance. Encouragement given to
songbirds, or to those birds especially given to devouring insects, became a practical consideration. Birds were tempted in a variety of
ways. In a large Albany garden at the end of the eighteenth century, the
skulls of animals (horses and oxen) were fixed to the tops of fence
posts, to be used as birdhouses. Elsewhere, conveniently wide eaves
were constructed inside porches, and charming miniature houses were

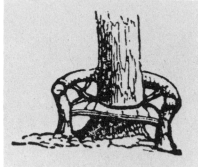

Several of Downing's "embellishments" from *Landscape Gardening*.

placed on pillars in gardens for the same purpose. Fencing became an embellishment, although still necessary for the exclusion of animals. In towns where mansions rose to three stories and had elaborate accents on their roofs of balustrades set like tiaras, the fences along the street would be made recognizably to match roof designs.

In the nineteenth century, space became a challenge to be filled. Simplicity equaled sadness. A stretch of lawn needed flower beds and shrubberies and sites for different sorts of trees to be grouped in clumps. With the advent of cast iron in the second quarter of the century, large urns were available to serve as ornaments in themselves, not requiring planting until much later in the century. Iron dogs could lie about. Iron deer could watch for passersby on the street. And best of all, iron furniture could take the place of rustic benches while still appearing very like them, with bare limbs and branches seeming to intertwine. As the possibilities of cast-iron furniture grew, grapes (as seen in festoons on the tops of Victorian sofas and chairs) moved out to grace sets of garden furniture; they encrusted the backs of upright little iron chairs gathered around similarly vine-entwined, grape-bearing tables. Hitching posts sported finials of cast-iron horses' heads, to indicate their function, or little cast-iron boys, usually "appealingly" in rags, though some wore uniforms, who held out a ring to hold a bridle chain.

Summerhouses became equally embellished, far beyond, one can only believe, ordinary comfort. Repose is not induced by shells stuck into a plaster ceiling overhead or by contemplation of views through tangled limbs. Still, there was something for all tastes. After Parmentier raised a rustic summerhouse to a height requiring a staircase under it, in order to command a view of New York from Brooklyn, and Downing enthusiastically approved the idea, rustic lookouts proliferated to a degree that brought protests and their subsidence to levels appropriate to their surroundings—high in public parks but low in private gardens. And yet we know that Longfellow, and undoubtedly others, effected sitting-out places in large trees. Throughout the nineteenth century, summerhouses—formal and latticed, or rustic and picturesque—held their own, as American as could be.[1]

As American home building spread rapidly toward the West in the nineteenth century, fencing became a consideration of first importance. Originally a necessity, it developed decorative overtones. In the cities for obvious reasons, cast-iron fencing became popular, and in the South it flowered into balconies across house-fronts like rows of iron lace. In older towns and suburbs, wooden fencing in rails, spindles, boards, and pickets continued serene, although it, too, began to have fanciful variations. After an elaborately ornamental stretch of fence along the front of the house, the rest of the property might seem to require different sorts of fencing on the sides and back. The author grew up in a garden with a nineteenth-century assortment of fences. A black cast-iron railing of plump, stubby banisters, not unlike those in the white wooden balustrades around the piazza and the roof, ran across the front of the house and the "front garden." From the corner on the side bordering a lane, there was a high solid fence, with a spruce hedge to hide it from the house. At its lower corner, where it turned to hide a house fronting on the lane and overlooking the front garden, it suddenly sported a pretty lattice top. As it continued down the side of the main garden, past the lawn and the flowerbeds and the pear orchard to the vegetable garden, it managed to hide a whole row of little houses with a neat but very high fence of painting palings. This section was easy to see through but hard to climb. At the very bottom of the garden, the fence became solid again, against a field and a millpond, and then repeated its variations up the other side toward the stable, trying to shut out another row of houses. This variety of fencing was not at all unusual. Children were used to coping with these different styles at their own homes and those of others. We loved the wooden fences and felt that the cast-iron front fence was ugly, although my childhood sorrow was that we did not have an iron dog under the tulip tree and copper beech.

By the middle of the nineteenth century American cast iron's popularity was assured. A new method of casting improved the quality of iron-made products, beyond the capacity of the blast-furnace method, and foundries sprang up in all areas of the country with convenient coal—in Pennsylvania, especially, the area from which the South imported its supplies. The new method is described in *A Dictionary of Arts, Manufactures, and Mines* by André Ure (New York, 1856):

> The operations of an iron foundry consist in remelting the pig iron of the blast furnace and giving it endless variety of forms by casting it in molds of different kinds prepared in appropriate manners. Coke is the only kind of fuel employed to effect the fusion of cast iron.

In an advertisement of 1787 in the *New York Daily Advertiser* the superiority of cast iron made in an "Air Furnace" over that from a blast furnace was said to consist in all the dross being expelled by a second melting, whereas the blast furnace's first melting of the crude ore leaves impurities that cause the ware to be rough and spongy. An added inducement to purchase American cast-iron products was the six-pound duty laid on foreign castings and the fact that the American articles were equally good.

Another advertisement in the *Perry Freeman,* December 6, 1885, gives an idea of the articles available from Robert Wood and Company of Philadelphia, flourishing manufacturers of ornamental cast iron.

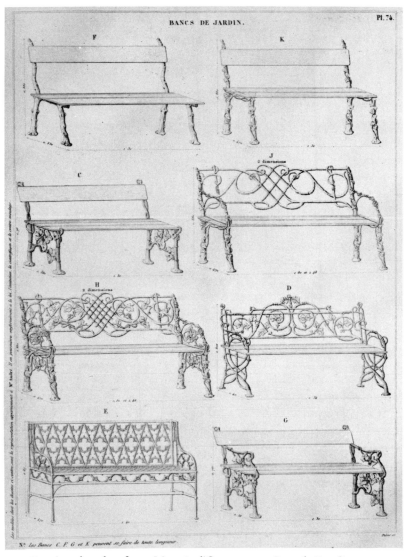

Cast-iron benches from *Magasin d'Ornements en Forte de Fer.* Courtesy
Essex Institute, Salem, Massachusetts.

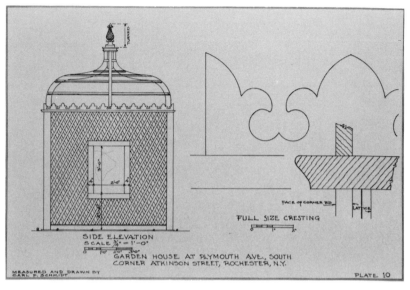

Garden House at Plymouth Avenue, south corner Atkinson Street, Rochester, New York. From Carl F. Schmidt, *Fences, Gates, and Garden Houses* (Rochester, N.Y., 1963).

Wood's Ornamental Ironworks

Ridge Avenue, Philadelphia. The attention of the public is invited to the extensive manufactory and wareroom of the subscriber who is prepared to furnish, at the shortest notice, Iron Railing, of every description for Cemeteries, Public and Private Buildings, also Verandahs, Fountains, Settees, Chairs, Lions, Dogs, etc. and other Ornamental Iron Work of a decorative character, all of which is executed with the express view of pleasing the taste, while they combine all the requisites of beauty and substantial construction.

With the advance of the American machine age and industrial production, gardeners and gardening were carried along, like all others and all else. With the perfecting of the manufacture of wire, both smooth and barbed, which was soon to enmesh the West, there was a flourishing trade in items like trellises, edging to surround flower beds, stands to hold collections of flower pots for parlor or conservatory use,

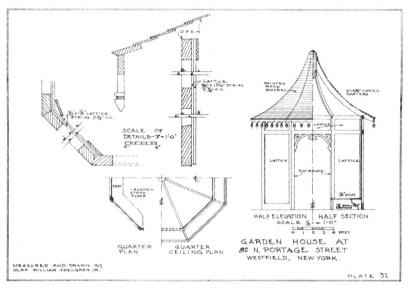

Garden House at 82 North Portage Street, Westfield, New York (measured
and drawn by Olaf William Shelgren, Jr). From Schmidt, *Fences, Gates,
and Garden Houses.*

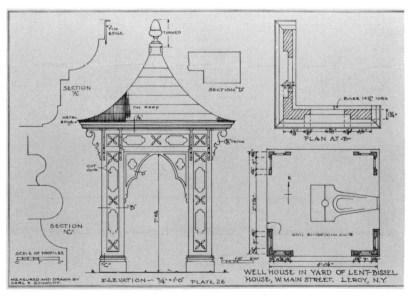

Wellhouse in yard of Lent–Bissel House, West Main Street, Leroy, New
York. From Schmidt, *Fences, Gates, and Garden Houses.*

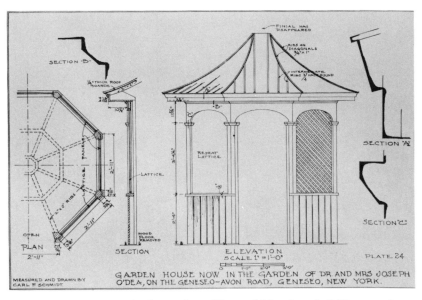

Garden house, now in the garden of Dr. and Mrs. Joseph O'Dea, on the Geneseo–Avon Road, Geneseo, New York. From Schmidt, *Fences, Gates, and Garden Houses*.

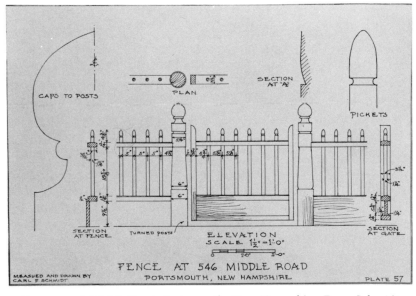

Fence at 546 Middle Road, Portsmouth, New Hampshire. From Schmidt, *Fences, Gates, and Garden Houses*.

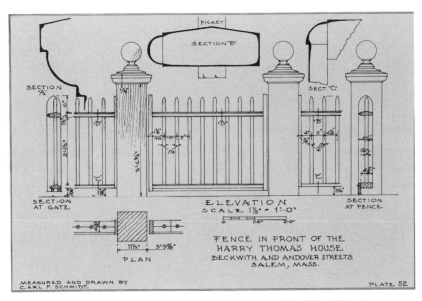

Fence in front of the Harry Thomas House, Beckwith and Andover Streets, Salem, Massachusetts. From Schmidt, *Fences, Gates, and Garden Houses*.

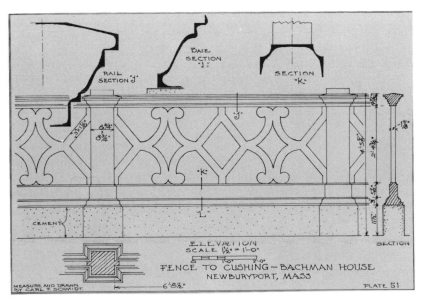

Fence at Cushing–Bachman House, Newburyport, Massachusetts. From Schmidt, *Fences, Gates, and Garden Houses*.

frames for ambitious excesses of flower arranging, and even chairs (where the use of additional cushions in the seats was advised).

We have included illustrations of many pretty suggestions for seats and one or two rustic summerhouses for celebrated gardens. In addition to these inspiring pictures, showing the befores and afters of embellishing the simplest garden structures, we are fortunate to have had an enthusiast for the actual carpentered structures still remaining from the nineteenth century. Carl F. Schmidt went about on "sketching and measuring jaunts" and left us with records of what really was there and directions for making accurate copies. He was generous with his sketches and with his concern for restoring the past correctly. He had an eye for the charming touches that transformed fences, seats, and summerhouses from the purely utilitarian to near works of art.

His book, *Fences, Gates, and Garden Houses,* published in Rochester in 1963, includes examples from the "Post-Colonial, Greek Revival and . . . Victorian style houses." We present a few of his instructive illustrations, made from his drawings and photographs. It is my great regret that this must represent my gratitude because, although he knew this was what I intended to do with his sketches and sent me more, it took me so long to write this book that my letter to ask final permission and show what I had done was, sadly, returned unopened.

These are what Mr. Hovey called "fanciful light summerhouses." The names and locations are sufficient comment on the designs to give us confidence that Mr. Schmidt's drawings are truly representative of the best in nineteenth-century garden structures, at least in areas where wood was abundant. Away from forested areas, construction consisted of locally available materials—brick, stone, concrete, tabby, and finally cast iron. Basic goals remained the same: shelter, shade, privacy, outlook, beautification, and the disguising of functional uses.

And, of course, there would be a reflection of regional taste. Little Greek temples on mounts would please some parts of the country—as see the Howards' garden in Toronto. In more extended southern plantation plans—as in Rosedown Plantation near New Orleans—estate and doctors' offices would be housed in pretty little buildings used as accents in the general layout, focal points at the ends of long, ornately planted walks through garden beds. In Tennessee, Andrew Jackson, like Washington at Mount Vernon the century before, incorporated a

"Pleasures of the Country—The Sweet Home" (Currier and Ives). Note the rustic bench and the cross-barred gate, the picket fence, and the gateposts. Courtesy the Collection of Heritage Plantation of Sandwich, Massachusetts.

"necessary" into his formal garden plan. Somewhat centralized ornaments like obelisks and sundials often served to define the garden's design. And of course there were many popular "embellishments," to adopt Downing's word for them, in the form of seats and benches conveniently placed for observing landscape features considered particularly impressive or charming.

Within the reach of small householders, but also in great favor when scattered about large estates, were all sorts of ornamental urns and baskets. An empty urn upon a pedestal was considered sufficient in itself until joys were revealed in filling it to running over with exotic plant material. Rustic baskets abounded near houses, simply sitting upon the lawn, although Scott fancied hanging them exactly in the line of vision from porches and arbors.

There must have been many variations on the urge to fill space to "beautify" it, which could lead to later whimsies like old rowboats full of geraniums on front lawns or tripods of birch saplings supporting kettles full of petunias over imaginary flames. Fortunately for restorers today, a touch or token of all these efforts to ornament and embellish will suffice to evoke the spirit of an age when space had to be filled to avoid seeming "sad."

NOTE TO CHAPTER 15

1   Here lies a point to be made, however gently, to those who have adopted new words for old bottles. In a recently reprinted 1811 *Dictionary of the Vulgar Tongue, gazebo* is defined as "an elevated observatory or summer house." In the latest edition of the Oxford English Dictionary, the definition of *gaz-ī-bo* seeks to convey that the word may be slightly spurious, of East Indian origin but coined by someone who knew Latin and made a little jest, as one would say *video,* but approximated it with *gazebo,* to imply a place where one could gaze out all around. The word connotes a raised lookout rather than a shelter on a lawn. It does not seem to have been a name used, even in jest, in North America until modern times.

SIXTEEN

Bedding Out

A T midcentury, a whole new
style of gardening arrived from abroad, with such impact that it
marked the century as its own and lasted, in certain aspects, well
into the next century. After the early-nineteenth-century enthusiasm
for natural, as opposed to geometric, landscape design, it is strange
that this new orientation should be the fantastic practice known as
"bedding-out."

For years, American gardeners had been happy to nurse their special
favorites in borders on the edges of their lawns or orchards or vegetable gardens, or in beds cut into advantageous positions to be enjoyed
from the house. Whole gardens had been laid out to be walked in,
among beds of assorted flowers and shrubs often trimmed along the
edges with bricks, stones, and miniature hedges of clipped box.

Instead of the dignity and grandeur of curved drives and paths with
trees and shrubs in clumps leading to vistas, instead of wide borders of
graded flower plantings to hide boundary lines, the new fashion advocated plantings that imitated Oriental carpets, ribbons, and spouting
fountains, all worked out in masses of vibrant bloom and variegated
leaves. Lawns were to be perforated with sudden random plantings of
gigantic single plants or made to sport a layout of low plantings. Lines
of identical small plants flanked by lines of other plants of a contrasting

color were laid out beside paths and drives like colored ribbons. This new approach reflected the suburban desire to impress passing strangers, even at the sacrifice of domestic privacy and comfort. It persists in the placement of iron statues (of deer or dogs, for example) on front lawns—facing the street.

Such a departure from established practices stemmed from the simultaneous occurrence of two seemingly unrelated events abroad: the lifting of the tax on glass in England and the influx there of horticultural material from the North American Southwest and the South American tropics. Cold-frame and greenhouse culture became popular for all social classes, and vast supplies of small, brilliant "annuals" became available.

During the Civil War, little attention was paid to developments in domestic horticulture, but the imported "modern style" caught on after the war's end. It was spurred by the numbers of small landholders who, despite confined garden spaces, could exhibit their taste for color and variety, and for startling effects. The chief advocate of the new aesthetic in the United States rose to prominence in the third quarter of the nineteenth century, just as its practice began to fade abroad.

Peter Henderson, of Jersey Heights, New Jersey, published *Gardening for Profit,* on growing vegetables, for market gardeners; *Practical Floriculture* for commercial florists; and *Handbook of Plants* for nurserymen. In *Practical Floriculture,* a small book of about 250 pages, he treated a wide range of subjects.[1] Referring to flower beds, he announced he had found nothing to surpass planting in ribbon lines for a truly grand effect, especially as done at the Crystal Palace in London and the Jardins des Plantes in Paris. The response from his readers was so great, he said, that he later was impelled to write a book for American amateur gardeners, using the "plainest language" of which he was "capable."

Henderson's *Gardening for Pleasure,* which appeared in 1875, eased us amateurs into a new garden concept.[2] He leaves the planting of lawns with trees and shrubs to be covered by landscape gardeners like Downing and Scott and limits himself to flower beds.

> Old fashioned mixed borders of four to six feet wide along the walks of the fruit or vegetable garden were usually planted with hardy herbaceous plants, the tall growing at the back with the

lower growing sorts in front. These, when there was a good collection, gave a bloom of varied color throughout the entire growing season. But the more modern style of flower borders has quite displaced such collections and they are now but little seen, unless in old gardens or botanical collections. . . . Then again we have the mixed borders of bedding plants, a heterogeneous grouping of all kinds of tropical plants, still holding to the plan of either placing the highest at the back of the border if it has only one walk, or, if a bed has a walk on each side, the highest in the middle.

The "mixed system," he says, "has advocates who deprecate" the modern plan of massing in color as being too formal and too unnatural a way to dispose of flowers. *Mixed* was opposed to *massing*. "Ribbon," apparently, is massing strung out. A succession of solid colors in bands, massed in a geometric design, was ideal.

As examples of "modern" methods, Henderson gives us two plans with centers five feet high, to be used as a large feature in the middle of a lawn. He begins with a design of ten concentric circles, the dimensions necessarily somewhat large. In the center, we place the tallest of the appropriate plants, remembering that all will need to be "pinched back" while growing so the "outline will form a regular shape from the center or highest point down to the front." He gives us two lists of plants to choose between. From the center out, the first includes (his spelling):

1. *Canna indica zebrina,* with striped leaves
2. *Salvia splendens,* with scarlet flowers
3. *Golden coleus,* with orange and brown leaves
4. *Achyranthes lindeni,* with crimson leaves
5. *Phalaris arundinacea var.,* with white and green leaves
6. *Achyranthes gilsoni,* with carmine leaves
7. *Bronze geranium,* with golden bronze leaves
8. *Centaurea candida,* with white leaves
9. *Alternanthera latifolia,* with crimson and yellow leaves
10. *Lobelia paxtoni,* with blue flowers

The second list is similar but contains touches of chocolate and gray in the leaves and allows a ring of delphiniums and one of nasturtiums, purely for their blossoms. These schemes would give a handsome effect inside the circular drive of a front entrance.

Henderson introduces many new annuals with the term *tropical*—a loaded word, which implies the plants will not stand cold weather. Let that be a clue, and a warning—for the "modern" plan, with its thousands of plants, will disappear with the first touch of frost. When we prepare to fling our floral rug upon the lawn for thousands to admire, we must know that this generous extravaganza will have to be repeated meticulously every spring. To maintain even a small floral rug—or clock, or map—one needs undivided attention and a glass house with a stove, or a greenhouse, through four seasons.

Henderson is not disposed to argue with detractors, however. His visits abroad, in 1872 and 1885, convinced him of the merits of carpet bedding. He says that the "carpet styles" of massing plants, as done in public gardens in London, were *interesting to the people* (emphasis mine) in a way that no mixed border could ever be. Nobody can deny, though many may regret, the power of the mob, however gentle. And there we have the revolution in garden styles that swept and still sweeps areas visited by people in crowds. Henderson scorns all efforts not along the same lines in the United States (with the exception of outstanding successes in Chicago, Philadelphia, Boston, and Allegheny City). He regrets that Central Park seems uninterested in this form of gardening due, he feels, to "incompetency or want of taste" and rejoices that several large private grounds near New York City have been "noted for years for their grand display of carpet bedding—unequaled perhaps by anything else in the world."

The large number of plants required for this type of display must be set out to form continuous masses shortly after planting. Four thousand plants are required to cover an area of one thousand square feet. In 1886 Mr. Hoey's garden at Long Branch, New Jersey, contained one and a half million plants in four beds, "so artistically" arranged that at a distance they might be mistaken for carpets laid out to air on the lawn. In fact, an old farmer and his wife, passing by in a rainstorm, drove in to warn the servants to get the carpets in before they were ruined. Mr. Hoey's efforts "in clothing his grounds in all this gorgeous coloring" were rewarded; tens of thousands viewed his garden annually, and he had "more to do with extending the taste for lawn decoration of flower beds than perhaps all other sources combined!"

The carpet style requires plants that can be kept down to a few inches above lawn level. Appropriate plants include succulents like echeverias, sedums, and mesembryanthemums, combined with low-growing alpines like ajugas, cerastiums, lysimachias, lobelias, ivies, and so on. The method of planting admits an infinite variety in forms and "contrasts of color."

So important were the exact tints of the leading flowers and the colored foliage (leaves must also contrast), decided on before planting, that dealers abroad supplied colored papers to be followed. And, though contrasting flower colors and variegated foliage are desirable, parts of the design were separated by "well defined portions of turf," to accentuate the colors. Henderson says the whole has a much better effect if a liberal amount of green is introduced. Even the most skillful scythers and sicklers would have had a hard time among the medallions and arabesques; the elegantly top-hatted man with the reel mower, seen in the preview, was in his element.

Peter Henderson's second edition included some sure-fire plans and patterns for fancy flower beds within the amateur's reach.

1. A simple design, around a center as of a four-petaled flower superimposed upon another with less-defined separated petals. This requires only two sorts of a single plant: coleus, crimson and golden.

2. Another simple floral design, of six almost round petals around a center. The plant featured in the center is crimson coleus (*Coleus verschaffeltii*). The petals are alternated in scarlet geranium General Grant and pink geranium Queen Olga.

3. A very fancy design, of eight foliated points around a large circle, with a solid center of the rainbow plant (*Alternanthera paronychioides major*) ringed by *Alternanthera aurea nana*, obviously golden. The foliated points are alternately dwarf scarlet nasturtiums and blue lobelia.

4. A design for a flower bed of generous, carpetlike dimensions with a floral design inside a square. The center is our scarlet geranium General Grant, ringed narrowly by the fountain plant (*Dracaena indivisa*) and widely by pink geraniums, this time unnamed. A narrow ring of *Anthemis coronaria*, or double yellow marguerite, finishes the center and leads to the base of the petals, which are all

1. Blue Lobelia
2. Alternanthera aurea nana, yellow
3. Achyranthes Lindenii, deep crimson
4. Geranium Mountain of Snow, white
5. Anthemis coronaria, Double yellow
   Marguerite

6. Pink geranium
7. Dracaena indivisa, or fountain plant
8. Geranium General Grant, scarlet

Flower bed for a private garden. From Henderson's *Gardening for Pleasure*.

of white geranium Mountain Snow. These appear to be resting on a solid deep crimson background of *Achyranthes Lindenii*, bordered geometrically by yellow *Alternanthera aurea nana* and finished off by a wide border of blue lobelia. Massing in colors at its best.

Having been initiated by Peter Henderson into the more difficult manifestations of Victorian bedding-out, we can consider the simpler

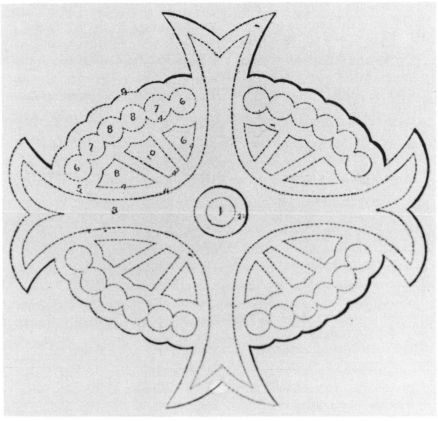

1. Vase	6. Geranium, Wonderful
2. Coleus, Bacon	7. Geranium, Madame Thebaud
3. Coleus, South Park Gem	8. Geranium, Mountain of Snow
4. Achyranthes metallica	9. Gnaphalium lanatum
5. Coleus, Mary Stewart	10. Coleus verschaffeltii

Flower bed for a public garden: "Fancy Bedding at Lincoln Park, Chicago." From Henderson's *Gardening for Pleasure*.

and more feasible effects of ribbon bedding. The plants used here can be cheaper and of more vigorous growth, because they are set apart and do not need as much pinching back to keep the whole expanse level. Again, however, "sharply contrasting colors make the most striking effects: and plants must be trimmed to prevent overlapping and to form 'clear defined lines.'" Two or more lines of color make "ribbon lines" to go along the margins of drives or walks. "Massing in colors" contributes to the designs given, laid out by George A. Solly and Son, a firm in Springfield, Massachusetts.

At about this time the voice of William Robinson was raised in England. In 1883—between Henderson's admiring visits—the first edition of *The English Flower Garden* appeared, with an opening shot against "bedding-out or marshalling the flowers in stiff lines and geometric patterns" as "entirely a thing of our own precious time." Robinson declares carpet gardening "simply a further remove in ugliness."

Henderson, if he hears, gives no sign. Kindly, for those who may have a lawn but are not up to laying out the living carpet, he has simpler suggestions. He suggests cutting holes into the lawn at intervals and planting in each a "single specimen of stately habit." He says, for example, that varieties of the castor-oil plant will grow ten to twelve feet high in one season, "a particularly striking plant." Or one might plant a group of scarlet salvia six feet high, its dazzling color contrasting superbly against the green of the lawn. Or take a plant with colored foliage, like *Amaranthus tricolor gigantea* (Joseph's coat), which grows to six feet, with leaves that "exceed in brilliance of color anything we know of in foliage: scarlet, crimson and gold predominating." And, of course, "more somber" tints may be used. Pampas grass attains a height of six to ten feet and has a graceful appearance. Other ornamental grasses, like the Japanese ribbon grass and the zebra grass, have useful flower spikes, which when dried last for years and make "unique parlor ornaments."

In the late twentieth century, one may query, "Did they really *do* all those things?" They did. Witness today the parks surrounding Niagara Falls (particularly on the Canadian side) and the plantings of the Boston Public Garden in Massachusetts—Victorian bedding-out, a sight that must be perpetuated to be believed.

NOTES TO CHAPTER 16

1 From laying out "ornamental grounds" and flower beds through propagation, pest control, greenhouses and window gardening, and on to a gardener's "diary of operations" and plant lists.

2 We shall follow the first edition, but later editions continue its enthusiasm for bedding-out, in designs both floral and geometric, and for "massing in colors," which was Henderson's special delight.

SEVENTEEN

Neighboring Improvements

wo features of New World
gardening and landscaping are peculiarly American: the suburban custom of uniting the front lawns of however many houses there may be
on both sides of a street to present an untroubled aspect of expansive
green to the passerby, and the—later-adopted but no less prevalent—
twentieth-century custom of surrounding the lower sections of suburban houses with evergreen shrubs, quite simply called "foundation
planting." In their inceptions, these practices are a hundred years apart,
but both are followed today, almost as if by law, and both are uniquely
American. Their development can be understood by tracing thoughts
on small-scale gardening and landscaping in the new country.

The many books on landscape gardening in the first half of the nineteenth century, both those imported from England and those published in the United States, generally agreed on matters of basic design
and its execution in plant materials. Plans for the flower garden were
compact arrangements of squares, at first sight similar to designs for
vegetable gardens, although beds in the vegetable garden were laid out
in rows and those for flower gardens showed borders around central
features. Trees and shrubs were arranged, as in the late eighteenth century, in "clumps." The drive curved. To overcome the criticism that
drives seemed to wind for winding's sake, groups of trees or shrubs
were placed so as to appear to be the reasons for the curves. On the

whole, designs assumed a spaciousness, a prominence of position, access to long views, and the possible blessings of water, in sheets or streams.

IN his *Art of Beautifying Suburban Home Grounds,* Frank J. Scott addressed a new sort of householder. He was concerned with the "millions of America's busy men whose daily business away from home is a necessity" but who yet desire around their homes "the greatest amount of beauty which their means will enable them to maintain" and who appreciate "all the heart's cheer, the refined pleasures, and the beauty which should attach to a suburban home."

Scott acknowledged his indebtedness to his friend and instructor, A. J. Downing, by dedicating the book to him. Published in 1870, nearly twenty years after Downing's sudden death, the book paid tribute to its origins. In a gracious introduction, Scott defined his objective as the furthering of Downing's intentions as expressed in his latest books on cottage residences and country homes—in short, to adapt "decorative gardening" to the "small grounds of most suburban homes." No book had previously been devoted to "suburban home embellishment." Landscape gardening, Scott said, is a term misapplied "when used in connection with the improvement of a few roods of suburban ground."

Scott did not intend to treat landscape gardening on the scale or with the thoroughness of the English masters of the art or of Downing. He remarked with modest sincerity that the American ability to design on a grand scale is shown in the cemeteries ("renowned even in Europe for their tasteful keeping") and in that wonderful creation, New York's Central Park. (There is "illustrated the power of public money in the hands of tasteful genius to reproduce, as if by magic, the gardening glories of older lands.")

But public parks, Scott observed, are not substitutes for beautiful homes. He saw a need to "epitomise and Americanize" the principles of decorative gardening so that even the half-acre of a suburban cottage, "if the house itself is what it should be, may be as perfect a work of art . . . as any part of . . . Chatsworth." He noted that, compared with the English, Americans were yet novices in the art of gardening and that the "exquisite rural taste" found even among the poorer

"All the heart's cheer, the refined pleasures, and the beauty which should
attach to a suburban home."
(All illustrations in this chapter are from Frank J. Scott, *The Art of Beautifying Suburban Home Grounds* [New York: Appleton, 1870].)

classes in England, which inspired the praise of Washington Irving
thirty years earlier, was still far ahead of the American awareness. He
noted that John Claudius Loudon's works were comprehensive and
useful but "too voluminous, too thorough, too English to meet the
needs of American suburban life." Loudon's successor, Edward Kemp,
in his "complete little volume entitled *How to Lay Out a Garden* [had]
condensed all that is most essential on the subject *for England*."

Scott pointed out that American homes differ so widely in climate

A tree with "expression."

and domestic arrangements from those of the English that plans cannot readily be adapted to the new circumstances. Even the thoroughness of British arrangements of outbuildings for the housing of
maidservants, manservants, and domestic animals had to be condensed
or dispensed with in America as unneeded, due to the great cost
of labor.

Scott began to rectify the omission of appropriate guidance for the
new territories by first covering some basic information on landscape
design, usually glossed over in other works. He observed that "un-

scientific lovers of nature and rural art" must learn to know their trees and shrubs. *Familiarity with the materials* is all-important. A key word for the individual characteristics of each horticultural specimen to be used is *expression,* which Scott italicized. The planner who would create home beauty must be aware of the *expression* of trees and shrubs, "as produced by their sizes, forms, colors, leaves, flowers and general structure, quite independent of their characteristics as noted by the botanist." Like George Marsh, he saw trees as possessors of personality. The relative importance of each of the subjects in his 600-page book becomes readily apparent. The first half deals with the theory and mechanics of gardening;[1] the other half is completely devoted to "Trees, Shrubs, and Vines." In the Appendix, the list of materials other than flowers available to the American gardener in the mid-nineteenth century follows Scott's advice, with added comments by other writers.

Scott was first to deal, at length and in digging depth, with a phenomenon new both to home landscaping and to city planning: the individual's responsibility to neighbors. Fanny Kemble criticized as unseemly the variety in American homes, due to each householder's apparently having designed his own house. She preferred the elegant homogeneity of England, where whole blocks of the same style in brick and stone could be divided into individual portions, built in pairs, or allowed to stand separately though still harmoniously. In contrast, books on architecture urged the American homeowner to think for himself within the limits of "taste" and "refinement"; only to these goals should a dwelling submit, even if this required that it be covered in vines.

Scott's point was that, given a "suitable" house, grounds could be made beautiful, whether they consist of half an acre or five acres, and beautiful not merely to the owner but to everyone around. He advocates, wistfully, the formation of "companies of congenial gentlemen to buy land enough for all" (of them) in a promising locality.

Where the ground permits, suburban lots can have frontages of one, two, or three hundred feet each, and extend back to a depth (when possible) of four times the frontage. These proportions Scott considers the best "for improvement in connection with adjoining neighbors," in addition to insuring the blessings of nearby friends and good walks

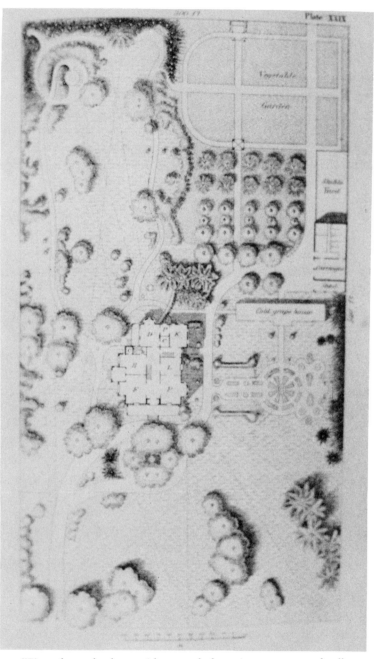

"First–class suburban residence and plantation on a corner lot," exposing all its front yard for public enjoyment.

and roads. He saw that cities would become hives for working, their suburbs like villages, though of more cosmopolitan and generous character than old-fashioned villages. He saw this "half-country, half-town life" as "the happy medium" and as a realizable ideal for many Americans, meeting, more than in any other form, the wants of cultivated people. Lots of between half an acre and five acres are large enough to afford the finer pleasures of rural life, and the wives and daughters of businessmen will no longer have to contend with the isolation of farms.

Of course, there are responsibilities. "*The beauty obtained by throwing front grounds open together* [emphasis mine] enriches all who take part in the exchange and makes no man poorer."

> As a business matter it is simply stupid to shut out voluntarily a pleasant lookout through a neighbor's ornamental grounds. . . . no gentleman would hesitate to make return for the privilege by arranging his own ground so as to give the neighbor equally pleasing vistas to or across it.

But this is not all: "it is unchristian to hedge from the sight of others the beauties of nature which it has been our good fortune to create or secure." All fences, high walls, hedge screens, and belts of trees are "unchristian and unneighborly."

It is refreshing to come upon Scott's staunchly New World spirit after so many of his predecessors have bent their knees to the grand succession of English authorities. In all fairness, he admits, "conventional forms of planting," which have come down from feudal times, may be necessary in gardens near cities, which are likely to be "in proximity to rude improvements and ruder people" and are not usually deliberately carried on in a spirit of selfishness.

To those dreading too much exposure to even the most congenial neighbors, Scott says that only "front yards" are to be shared, not the domestic offices, which must be kept private and distinct. Should neighbors fear boisterous children running over the contiguous lawns, the children's families should train them to stop at a boundary line— say, at a thread drawn between the properties, or a tiny ditch sunk in the grass.

Landscaping of the back and sides may be determined by each house, in order to give foremost satisfaction to those looking out from

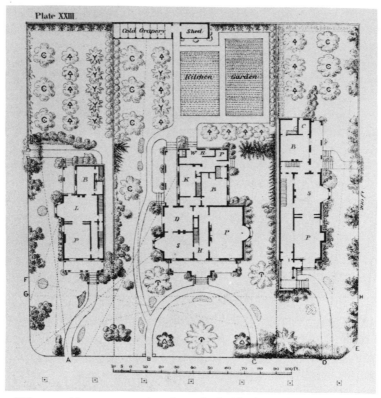

"Three residences occupying the end of a block 200 feet in width on lots 200 feet deep," a plan of several lots whose intermingling front lawns it would be "unchristian" to hide from neighbors; notice open front gardens and private developments in rear.

the main windows or verandas. Here the lawn must be made to seem as extensive as possible. Vistas can be created, with larger trees at the ends and shorter trees or shrubs closer to the viewer. Each window should have a different view, but on a small lot that may be asking too much. Instead of trees, shrubs and a few well-placed flower beds can provide interest. Scott suggests that the planner think of the lawn as a lady's gown of green velvet and of the flower beds as bits of lace or decoration. Too much decoration, ill placed, destroys the refinement of the gown. Finis.

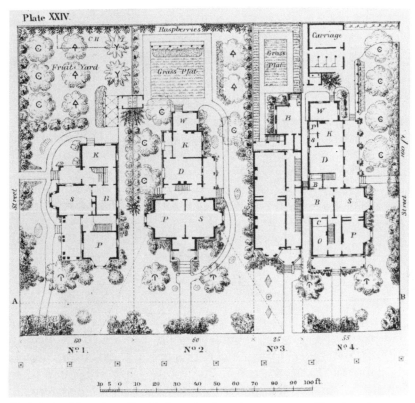

Plate XXIV.

Varied-sized lots, but all with front landscaping open to public view; lots are 150 feet deep, and there are four houses of differing styles.

Here is where the importance of *expression* in trees and shrubs becomes obvious. If the expanse is small, though covered with a fine growth of lawn, it must not be cut up or overcrowded with plantings. Large trees must be placed at the outermost boundaries, with smaller trees or shrubs in front of them. Flowers can be confined to borders or beds, near walks or gates or steps. Specimen trees must be chosen with their "expression" in mind.

He illustrates the elements of expression, first by considering shapes— pointed, rounded, flat-topped, fountain-shaped, and so on—and then by considering the plants' potentials for shelter, comfort, amusing eccentricity, simple beauty, grandeur, and other qualities. Size must be

An idea for a rose bed.

estimated at full growth and planting plans drawn to arrange for the space needed. Odd leaves, autumn coloring, contorted limbs, density of shade, and winter charm are all to be considered in making choices.

By the time homeowners have studied the whole range of recommended trees and shrubs, they will be able to prove their superior taste and refinement to the rear of their houses while exhibiting the proper spirit of "neighboring improvement" for all to see in their conjoining "front yards."

By "joining neighborhood improvements," Scott shows how the beauties of old and new places can be equalized. For instance, he asks us to consider three contiguous lots: The first, belonging to *A,* is in an "old village style," with big cherry trees, maples, lilacs, spruce trees, roses, and annuals; *B* is just moving onto a bare lot, on which he will build a "modern house"; *C*'s grounds have a growth of noble old trees, which "invited a house to make its home there."

So what is *B* to do? *A*'s yard is cluttered with a valuable accumulation of years, nothing showing to advantage; close turf mown to make a lawn is almost impossible. *B* therefore concentrates on a fine lawn, constantly cutting and rolling it until it is a sheet of green velvet. "Cut

How to make a "foot walk gateway" feature with two trained trees (the result is too explosively threatening to contemplate, but the plan itself makes a lovely design).

in the lawn, here and there near his walks, small beds for low and brilliant flowers may sparkle with sunny gaiety." At intersections, "low broadtop vases, rustic or classic as the character of the house may require, may be filled with graceful and brilliant plants."

In two or three years, says Scott, *B* will have the most charming place of the three. By contrast with his neighbor's crowded yard on one side and the forest on the other, his will be the envy of the neighborhood.

The morals of this are confusing, but we can see the origins of American suburban standards of a hundred years later, still unique in the world, which, outside America, prefers to sacrifice no space for public, let alone neighborly, enjoyment. Even Loudon's recommendation for his suburban villas—to vary their single ornamental specimen trees, each in the middle of its separate front lawn, by observing those chosen by neighbors and planting another variety—does not come near to Scott's ideal of generous contiguity.

NOTE TO CHAPTER 17

1 The first half contains essays on art and nature; decorative planting; what kind of home grounds suit businessmen; suburban compared with country places; sites and ground surfaces; dwellings, outbuildings, and fences; neighboring improvements; materials used in decorative planting; faults to be avoided; walks, roads, and arrangements; the relative importance of lawn, trees, shrubs, and flowers; care of the lawn; artificial adaptation of trees and shrubs; plans; renovation of old places; flowers and bedding plants; drainage and cultivation.

EIGHTEEN

How to Make Country Places

N his *Supplement* to the sixth edition of Andrew Jackson Downing's influential *Landscape Gardening*, Henry Winthrop Sargent has endeavored to follow step by step in his neighbor's and preceptor's footsteps. One step brings him to Central Park; several more steps get him through forests of ornamental trees and shrubs, deciduous and evergreen. He imitates Downing's original format, with a chapter on country places, visited and referred to as "History," and another chapter entitled "General Remarks." We will follow him now through "How to Make a Country Place," which involves his own experience in creating Wodenethe on the Hudson, described briefly by Downing as a "bijou"—perhaps even before Sargent really extended himself.

Sargent says that two sorts of new places are most commonly attempted in the United States: a place without any foliage where all effects are produced by the spade; and a dense wood where all is accomplished by the ax. He illustrates the first with a history of his own place and the second by a description of Wellesley, the Hunnewell residence near Boston. Sargent has been familiar with each establishment since its commencement and is familiar with its motives and designs and the reason why every tree was planted or cut down. Because one was a dense wood and the other a naked field, they provide the best examples at his command of the two styles.

WHEN Wodenethe was purchased in 1840, it consisted of a partially built house in the midst of a wood, without any view, although there was a valley on one side, a range of mountains on another, and the Hudson River on a third. The fourth side gave onto a long range of country. The attractive points toward river, valley, and mountain needed to be opened up, and the houses scattered around needed plantings. All of the trees were tall and spindling, "hiding with their heads what they should not have concealed, and opening through their naked stems what ought to have been hid."

In order to form groups of trees, to hide objectionable houses and produce effects of light and shade (the basis of ornamental planting), Sargent topped those tall trees "with enough vitality in their lower branches to carry on the circulation." Seventy-foot trees reduced to thirty feet formed a background thicket, which, pruned and cut irregularly through the mass, yielded extended views, as seen in "View from Library Window."

Careful planting against masses produced a distinct view from each window and extended the vistas, making "as it were a series of cabinet pictures." An irregular plantation of the most ornamental trees was advanced in front of each window, to mask the masses through which the vistas were opened and to elongate the views. The "View from the Breakfast-room Window" shows that the ornamental facing of decapitated forest trees, along with rare trees in small irregular groups and singly, produces a harmonious effect. On the side of the house shown in "View across the Park," only the ax was needed to achieve a parklike effect, furnished by a group of native oaks, thinned out through the years to stand alone, as if they had been planted solely for this purpose.

On the boundaries of the place, trees were reduced to half or two-thirds of their height and, when they were thick and bushy, faced with ornamental plantings. An arboretum, a pinetum, and a walk encircling the whole were added. Spaces for vegetable and flower gardens and an orchard, as well as for all the ornamental plantings, had to be taken from the woods, the trees cut down and roots grubbed up.

WELLESLEY consists of about 200 acres, originally an unimproved portion of an old family place. Mr. Hunnewell selected forty acres of flat, sandy, arid plain for improvement in 1851. The area was covered

View from library window at Wodenethe, looking south. From Henry Winthrop Sargent's *Supplement to the Sixth Edition of "Landscape Gardening" by A.J. Downing* (1859).

View from breakfast room window at Wodenethe, looking west. From Sargent, *Supplement to ... "Landscape Gardening" by A.J. Downing.*

by a tangled growth of dwarf pitch pine, scrub oak, and birch, all of which had to be cut down and the land plowed up.

The first project involved the trenching over and preparation "with composted muck" of an acre or more for a nursery, where quantities of trees were planted, including Norway spruce, white pine, balsam, Austrian pine, Scotch fir, larch, beech, oak, elm, and maple. Most of these came from England and were not over twelve to fifteen inches high. A few native trees of greater age were added. A lawn was graded, subsoiled, manured, and enriched with compost, then cultivated for several years "to ameliorate and subdue the soil." The borders were trenched over, composted, and planted with a mixture of evergreens and ornamental trees. Potatoes were planted here annually until the trees grew enough to touch each other, and these yielded some remuneration for expenses.

The next project was to decide on the situation of the house and then to plant the avenues. On the Boston side, the entrance was planted, at one end, with alternating *Pinus excelsa* and *Magnolia tripetala* and, at the other end, with large masses of rhododendrons, *Kalmia latifolia,* mahonias, and other rare evergreen shrubs to form a frontage for a background of Norway spruces. As the road reaches the Italian garden, with a view of the lake on one side and the house and lawn on the other, the avenue effect ceases and an ornamental arrangement commences. On the other avenue, from the Natick side, the planting is all in double rows of white pine and larch.

Following this, the eight-acre lawn was planted with the best specimens obtainable from the nurseries and border plantations. Some large specimens were brought from twenty miles away and were added to form clumps, masses, and single specimens on the lawn, with each tree set into a cut circle, to be kept clean to the outer edges of drips from the branches.

Then the house was built, with, on one side, an extent of simple and dignified lawn, on the other "a French parterre or architectural garden," with fountains bordered by heavy balustrades, surmounted at intervals by vases. Steps lead through a series of terraces to the lake below, which is a fine sheet of water about a mile in extent, with varied outlines. From the French parterre, off on the right, stretch "the ornamental or English pleasure grounds." These form part of the same view and show a summerhouse artistically rusticated with colored-glass windows, which produce curious effects of contrasting colors in the stained glass.

Sargent waxes ecstatic while contemplating the Italian garden, "the most successful if not the only one as yet in the country." By moonlight—which he admits is an unusual test for any garden of that time—the views of the lake through the balustrades of the parapet and among the vases and statues that surmount it, with the sound of the splashing fountain and the unique-in-America features of formally clipped trees and topiary, "quite lead us to believe we are on the lake of Como."

And, indeed, this is a new criterion for judging American gardens—to feel oneself transported to another country. Its basis is the assumption that one will enjoy fragments of foreign architecture and gardens torn from their settings and resettled in a comparatively wild landscape. This was, in a way, a traditional pleasure of one sort of garden, where a scheme is imported and set down in an alien environment. For instance, where water and shade were scarce, as in Egypt and Persia and India, what could be more refreshing than a formal exhibition of a combination of these two features? In a rocky volcanic landscape, what more pleasing than architectural details, in terraces and balustrades? In a pleasantly dull agricultural landscape, what more exciting than the flinging down of an elaborate carpet of detailed compartments and embellishments, like Vaux-le-Vicomte and Versailles? Similarly, what a pleasing tour de force in Boston to feel oneself in Italy.

And topiary? The art held its fascination for some, despite all the eighteenth-century English wits and poets who decried the practice and despite the Abbé de Lille's long poem protesting the use of sheared hedges rather than softly waving limbs as backgrounds. To Hunnewell, Sargent says, "is due the merit of having first attempted to clip our white pine with the result of proving it able to bear the shears as well as hemlock or yew." The garden also contains successfully trimmed Norway spruces, balsam firs, arborvitae, English maples, beeches, and Scotch firs. Almost in expiation, it may seem to some, there is a walk through a wood planted to form a pinetum, full of the rarest and newest conifers and evergreen shrubs, which thrive exceedingly with the protection from the winter's sun.

After such unusual features, it seems unnecessary to add such pleasant items as vistas revealed through avenues of pine and beech, gardens for vegetables, fruits, and flowers enclosed in clipped hedges, and a steam engine to force water into a reservoir, to be piped out all over the garden. All this occurred within seven years, because Hunnewell, with taste and the ability to do things thoroughly and well, worked out a plan in which there was little to undo.

WE may now consider another sort of plan, advocated for a greater number of hopeful home builders at about the same time. A "plan of arrangement for a small village lot" was suggested in 1856 by the firm of Cleaveland, Backus, and Backus in *The Requirements of American Village Homes*. The house was placed at the center, twenty-five feet from the front, "compact and economical." It had a "good-sized parlor, a comfortable livingroom, an entrance hall large enough to serve as a sitting or eating room in summer and five bedrooms," all in a space of about twenty-seven feet square. A back kitchen, wood room, pantries, and so forth, were in an extension to the rear. Side covering was best applied vertically, although clapboarding would do. The cost was $1,500.

The plot was level, 75 feet in front, 150 feet deep, admittedly not large enough for "satisfactory cultivation in a general way." What the architects would consider "general" is hard to fathom, as they seemed to expect a great deal from this modest establishment. The veranda and parlor front windows looked toward the west. The hall and kitchen windows and the rear entrance faced south, the position "best adapted for comfort at all seasons of the year and all hours of the day."

A turfed lane wide enough to accommodate a load of hay ran straight along the northern line, separated from the property by a wire barrier on which a grapevine was trained. Two front gates opened onto paths about three and a half feet wide, which curved to meet in front of the veranda. From each gate, a path continued toward the back on each side of the house, meeting at the well and continuing toward outbuildings for a cow, a pig, and poultry, and on around the vegetable garden. Another path diverged on the south side, to lead in a reverse curve to the summerhouse and, around a planting of fruit trees, to the vegetables.

The front fence was "skirted by shrubs," but otherwise the semicircle was all grass. A suitable shade tree could go at each of the front corners. The space on the right—between the summerhouse and the street—was for a few fruit trees and a choice shrub or two. The space back of the summerhouse was for fruit only. On the northern side was a small flower bed of "fruitful" shape. Near the principal walk on the south side, several little beds "of various forms" were cut into the grass. Each bed was to be planted entirely with one of the following: petunias, verbenas, portulacas, violets, myrtles, or similar plants. A "creeping vine" went at the foot of each veranda post. A shrub shaded the downstairs bedroom window. Both the wellhouse and the summerhouse were "embowered." A thicket of evergreens hid the clothesyard, which bordered the cow yard and the vegetables (children could play in the clothesyard). Currants and raspberries hedged the vegetable garden. Trailing plants covered the stableyard fence. The manure pit was in a slope from the stableyard. The well was strategically placed for convenience of both house and stable.[1]

The owner of our vertically sided house in the midst of its plantings may not neglect the space in front and on the street. The outer edge of the sidewalk is the proper place for trees, judiciously selected and uncrowded. Space was also provided for a hitching post with a chain halter. After all this, the owner who set such an example, with all these improvements, need *say* nothing—his neighbors must be indifferent indeed if they could long resist such teaching.

DESCENDING in scale, though not importance, in American domestic architecture and landscaping, we may move from this plan for "a small village lot" to a more modest design directed at "the *intelligent cottager*" (author's emphasis) by Walter Elder. In *The Cottage Garden of America,* Elder laid down laws for both landlords and tenants in 1848 and 1854. Landlords, to make their tenants "comfortable," must select a healthful site on rising ground, away from stagnant water, and build a "convenient home," with a grape arbor attached to shade the back kitchens. A well should be built on the dividing line between each two cottages. Each lot should be fenced, manured, and trenched or plowed, and the garden should be laid out with gravel or tanbark walks, well edged. A patch near the house was to be sowed with grass for a "bleaching green," with a strong post in each corner to support a clothesline.

The garden, of an eighth of an acre, might have eight fruit trees, six currants and six raspberry bushes, four gooseberry bushes, fifty strawberry plants, six flowering shrubs, twelve roses, two grapevines, two "street shade trees," and twelve climbing plants. Any landlord who attended to the above "duties" would find his tenants "apt" to take care of the property and to be punctual with their rents.

The intelligent cottager might also build and own his own house—preferably one not more than one and a half stories tall, with a roof and coping on all sides. Gothic in style and twenty feet from the highway, with a fenced-in yard and the usual planting of two shade trees in front and a flower garden at the rear, it would have the same plantings seen in the landlord's layout for happy tenants. Loyally, Elder points to the existing gardens of Philadelphia as models and as proof that anyone may have his own house and garden.

NOTE TO CHAPTER 18

1   And where, we may wonder, was the outhouse, the earth closet, or the future water closet? Probably in the left rear corner of the extension. But why does nobody ever say, especially when the great J. C. Loudon so specifically dealt with this necessity in *The Villa Gardener?* In the second edition in 1850, edited by the faithful Mrs. Loudon, he discussed the earth closet of a small villa, probably to the rear of the villa and off the kitchen. There a superimposed water tank carried the waste down a drain into the manure pit by the stable at the extreme rear where, combined with raked-up leaves and discarded vegetable wastes, it could improve ordinary stable manure to make a valuable combination to be used in the gardens. I read this book at one sitting—or rather lying—in a hospital and was able to ask my visiting surgeon why these ardent gardeners and villa dwellers did not all die. "They did," he said. Perhaps this is why our own lovers of the common man and his overplanted houselot refrained from such encouragements. The wells between houses may have offered hazards enough.

Northern Lights

W E have seen the close rela-
tionship assumed by visitors to the early United States between the
areas on each side of Niagara Falls. To visit one without seeing the
other was unthinkable. There was also a feeling that Canada, from the
environs of the Great Lakes to the mouth of the St. Lawrence, held
much of the early history of the later United States. The settling of the
regions of Canada in the St. Lawrence valley and around the Great
Lakes was more unified and progressive than the less organized settling of the United States to the south. The Canadian settlers brought
with them their Victorian ideals of privacy and neatness, modest elabo-
rations of eaves and columns, compact planning of space against both
heat and cold, and a passion for gardening in however small a compass.
A simple elegance in brick, wood, slate, and stone took over the mid-
nineteenth-century Canadian landscape even as it was being cleared. A
simple sophistication overlaid the older, rough land clearing and the
timbered cabins and barns. Geometrical boundaries precluded expansive, winding, parklike features. With all the right angles of a great
grid already imposed, the idea of a geometrical garden must have
lacked charm. The desired norm included round beds of symmetrically planted annuals, a porch covered with blossoming vines, rose
beds cut into lawns groomed for croquet, and due attention to street-

Farm and residence of James M. Russell, Esq., Trafalgar. This and the following eight homesteads are taken from Canadian atlases of the period.

tree planting. Summerhouses, circular driveways, trellises, different styles of fences for different areas, and even pretty hitching posts occupied homeowners in an understated rivalry. Quality alone was significant.

The settlers' careful attention to the face of the land resulted in a sense that the land has been lovingly cared for, and the countryside maintains a beautiful, undulating, groomed, cheerful "face of nature" today. The same willingness to confine flower gardens to gay plantings by mailboxes on the highways and to small round beds near the house, or to neatly picket-fenced areas that also contain small orchards with neat paths, and to grow plants for fragrance and sentiment also marks the rolling landscapes of upper New York state, Ohio, and Wisconsin.

Canadian wildflowers had been world-famous since the times of the early Jesuits. Canadian trees had been brought out of the forests to preside over carefully trimmed lawns ever since Mrs. Simcoe came to help her husband govern Upper Canada and wrote glowing reports of their diversity and beauty. After stump fences had used the preponderance

Farm residence of W. C. Beaty, Esq.

of torn-up roots to restrain stock from wandering—just as New England's stone walls had grown from painfully cleared rocky fields—Canadian sawmills began to turn out picket fences and gates for front lawns, running board fences for the sides of gardens, and solid fences for private areas.

To critical visitors, the procedure of cutting down everything in sight and then replanting seemed violent and wasteful. Capt. Basil Hall, visiting the environs of Niagara Falls, related a "curious anecdote of landscape-gardening in America," possibly one of those stock stories (like the Indian servant who strained soup through his master's sock, or the Chinese boy who inflamed the Christmas pudding with kerosene) that traveled the outer reaches of empire and cropped up, like plantain, where the white man's foot had trod. Hall's account here is quoted from his contributions to Loudon's *Encyclopaedia of Gardening*.

A gentleman wishing to form a country residence . . . selected a certain spot in the midst of the wilderness which . . . might be converted, with a little trouble, from its wild state into a beautiful park, such as must have cost, in the ordinary process of old countries, at least one century, if not two, to bring to perfection. Some

Farm residence of John Breckon, Sr., Thornhill.

of the oaks and other trees were particularly beautiful and of immense size, and he determined on removing only those trees which encumbered the ground. . . . The trees were marked accordingly; but the proprietor was unfortunately obliged to be absent when the thinning took place, and the workmen, who from their infancy had known nothing about trees except that they ought to be cut down as soon as possible . . . decided that he had made a mistake, and that the small number of trees marked to be cut down were, in fact, those intended to be saved. The first thing, accordingly, that struck the master's eye on his return, was the whole of his noble grove lying flat upon the ground, while only a dozen or two craggy oaks, pines and hemlocks, destined for the fire, were left to tell the tale.

One of our pieces of great good luck in exploring nineteenth-century gardens lies in the reproductions of the "historical atlases." Among all the vital statistics and accounts of counties come line drawings of homes, factories, mills, stores, and hotels. With the amount of public exposure these drawings enjoyed, it is doubtful that the artist dared to be less than exact, although (as with Currier and Ives) the animals seem to possess a special vivacity and the owners a proclivity

Farm and residence of John S. Bessey, Esq., Georgetown, Ontario.

for croquet. The scenes and foregrounds, filled with barking dogs and with children rolling hoops, illustrate a careful regard for the layout of grounds. Formality sounds the keynote, and plants and shrubs are generously disposed. Driveways, paths, summerhouses, and ornamental trees are carefully placed, vines drape porches, flower beds cut into lawns, and orchards flourish nearby. An enormous assortment of fence styles enclose the grounds.

We have two sources for the sorts of plants to be found in the gardens. The first is a transcript of the diaries of John Howard, an architect who emigrated from England to Toronto in 1842, and the second is a book written by Mrs. Annie L. Jack.

OUR reference to John Howard's diaries draws upon his accounts of the years 1867 and 1868 in the gardens of Colborne Lodge, still standing (we hope as we go to press) in Toronto. These were two busy years, framed upon the maintenance of the house and gardens, beginning early each spring with the removal of snow from the glass roof and repairs to chimneys and the water-storage tank.

Repairs and improvements to the house and ground punctuate the

Farm and residence of Robert Haddow, Springdale.

diaries. The hotbed and glass houses need panes repaired or replaced. The eaves are painted, and the kitchen and scullery get new windows. The outside kitchen and basement rooms are whitewashed, the ice-house taken down, and a house for peafowl put up. Three stone-bordered beds are dug in the front of the garden, and the beds around the veranda are made up with stone. A new handle is put on the small rake for Mrs. Howard. During this time, the house was struck by lightning and the library flooded twice, which added to the series of chores only lightly sketched here. But always the gardening continued.

The winter saw the grapevines pruned and cuttings started in hot-beds. The snowdrops were up on April 10, when Howard tied up the sweetbriar. Seeds went into the hotbeds: zinnias, balsam, African mari-golds, red China asters, white and yellow chrysanthemums, blue lobel-ias, blue and white petunias, sweet scabious, China pink. Rings of sweet peas went around the trees on the front hill, by the basement windows, and were trained to run up the veranda. Hollyhocks were transplanted to the east side, near the apple and cherry trees. In mid-May nasturtiums and morning glories were sown. Howard had seed to spare: white and red candytuft, crimson stock, marvel of Peru, and

Residence of Sydney
Smith, Esq., Acton,
West Ontario.

common coreopsis. In late May, he divided and put out the "feather
plant" and spirea. Mrs. Howard recorded the different colors of hyacinths, and her husband trained the roses, honeysuckle, and sweet peas
around the veranda. "Wild white lilies" (*Trillium grandiflorum?*) were
brought in "from the bush," to go under trees near the hall door. Wall-
flowers, geraniums, sweet williams, and verbenas went into the beds.
Lady's slippers and red and white moccasin flowers were given away.
Sunflowers went on the left side of the lawn.

All this time, they sowed vegetables, enclosed the grapes on the
south front with glass to make them ripen, took up tulips, hyacinths,
crocuses, irises, and snowdrops, made wines and chopped down trees,
harvested apples and all else. And all cheerfully. Among all the stirring
accounts of the early settlers coping with the wilderness, none seems
to have worked harder or coped with more than the Howards.

OF the nearly 150 "horticulturists" listed in Liberty Hyde Bailey's
Standard Cyclopedia of Horticulture, 1950 edition, only one is a woman
and she is Canadian. Mrs. Annie L. Jack was born in Northampton-

Residence of George Cook, Esq., Esquesing, Ontario.

shire, England, in 1839, as Annie L. Hayr. She came to America in 1852 to attend a finishing school in Troy, New York, before going to teach school at Chateauquay Basin, Quebec, near Montreal. There she married Robert Jack, a noted fruit grower, and settled into Hillside Farm to grow small fruits and vegetables. A greenhouse for floriculture was added in time, and the place became a demonstration area for market gardeners. Her contributions to the press, published as "Garden Talks," described her own experiences and were extremely popular. Eventually, she published a handbook called *The Canadian Garden*. References here are drawn from the 1910 reprint of an earlier edition.

A poetic first chapter, called "Awaking," leads into a long list of subjects: Lay of the Land; Making a Hotbed; Causes of Failures; Transplanting; The Kitchen Garden; The Fruit Garden; The Pear; Plums, Cherries; "His Own Vine"; Red and Black Raspberries; Currants and Gooseberries; The Strawberry; Planning the Flower Garden; Lawn; The Rose; Bulbs for All Seasons; Border Plants to Choose From; Shrubs and Vine Hedges; Pruning and Grafting; Fighting Insects and

Residence of Dr. McGarvin, Acton,
West Ontario.

Residence of Robert Noble.

Diseases; Bordeaux Mixture; Annuals; Ornamental Trees; Window
and Cellar Plants; Envoi; Monthly Reminders.

We excerpt first a few paragraphs on planning the flower garden,
noting that the plan presupposes a lawn and an assortment of roses.

Planning the Flower Garden

How can you lay out your flower garden? . . . [T]here are no hard-
and-fast rules . . . and what suits one taste might clash with
another. A garden . . . shows as plainly as does the house the
characteristics of the residents. . . . The first thing, however, to
consider is the

A well-kept lawn, from Scott, *Beautifying Home Grounds*.

Lawn

for it is the canvas, and on it you paint with flower and shrub the picture that your fancy desires and your purse can gratify. . . . [I]t is not often one particular flower or shrub that forms the picture in the landscape, but the *tout ensemble*. So the lawn, however small, must be graded and ploughed or dug, harrowed, and then made level, for the more completely the soil is pulverized the quicker will there be a green sward.

There are varieties of grass mixtures advertised, but the foundation is best of Kentucky blue grass, three bushels to the acre, with a quart of white clover seed if it is approved and a little timothy seed to come up and cover the ground more quickly than the blue grass will do. . . .

If weeds come up they can be killed by the frequent use of the lawn mower. . . .

Mowing must be done so frequently that the grass can be left for mulch, and in autumn a covering of leaves is protective. . . .

The Flower Garden

South and south-east is a good aspect for the flower garden, and within sight of the house. If there are walks or drives, let us hope there are not too many curves or turns, for they cut up the grounds, make needless steps, and result in a footpath being often made across from point to point, in spite of, and to the annoyance of, the owner.

There must be shelter from high winds, and this can be given by groups of shrubs and ornamental trees. . . .

If it is suggested that beds in triangles and circles, with squares of geraniums and coleus, are not so much in evidence nowadays, we shall be referred to the city parks; but it is certainty that the fashion now in vogue, instead of this florid style of gardening, is long borders of choice perennials, summer-flowering bulbs and roots. Roses are set in beds, and ornamental shrubs blossom in glorious beauty and succession in the make-up of the floral picture.

Mrs. Jack covers the rose—the national flower—in careful detail. She notes the "old Persian yellow" and that Madame Plantier is the best of the double whites; that the York and Lancaster, "with its shadings from white to deep rose all on one spray," is "such a cheerful rose one forgives its flimsiness"; and that "the old Province rose . . . called the Cabbage rose" is "queen of country gardens." She observes that rambler roses have to be laid down in winter and that hybrid teas and Noisettes are too tender, unless kept in a pit and used for bedding-out. The hybrid perpetuals can be looked to for flowers—Margaret Dickson and Merveille de Lyon for whites, and Mrs. John Laing and Marie Bauman for pink and soft crimson. In Mrs. Jack's opinion, nothing can surpass a Jacqueminot. Earl Dufferin and Ulrich Brunner give a rich red, Paul Neyron a good pink. And nothing compares with Baroness Rothschild for pink, although it lacks perfume and is not entirely hardy.

For lilies, Mrs. Jack brings in the wild Canadian lily and combines the old Madonna with the old blue larkspur, among others, and she praises the soft yellow day lily. Dahlias and gladioli she finds capricious, doing better for cottagers than for some professionals.

We can depend, however, on the border plants, and her list is rich and rewarding.

Border Plants

The fashion of the day changes as much in gardening as in other things, and has brought to the front the dear old-fashioned flowers of English gardens, to which we, who are no longer young, look back with tender longing.

We have already discussed the border's bulbs, but when April comes we have as well the snow-white mass of the Arabis, the perennial candytuft, the daisy and moss pink. Later the forget-me-not, that should be in a clump in a half-shaded nook. By and by the *Alyssum saxatile* gleams out in yellow, and the Columbines shake their fairy bells in dainty colouring. The *Dielytra* [*Dicentra*] *spectabilis* gives its rosy hearts and fern-like foliage in sturdy strength and lengthened racemes that are always cheerful.

As the season progresses, the iris begins to bud and the lily of the valley suddenly shows its tiny bells. The latter should be planted in an out-of-the-way shady corner, for when among other plants it is apt to intrude, and after flowering no management can keep it presentable.

Then the peony asserts itself, and can make a garden a kaleidoscope of colour. Where there is a little space these plants can be set among tulips, and give a glowing fortnight of pink and crimson and cream, while the after foliage is pleasant to look on at all seasons.

The lovely old Canterbury Bell is the finest *Campanula*. *Gaillardias* are showy, but have a habit of sprawling all over the gound, yet not looking at home when tied up.

A clump of Iceland poppies look like glorified tissue paper, and open early in June, flowering, if seed is kept cut off, until October. But they are difficult to transplant, and a year of patience is required with seedlings, after which they keep on blooming continuously.

When the German, Spanish, and English Iris have done blooming, the Japanese takes their place like a bird of paradise. It requires a moist soil, and, being a gross feeder, the Iris will assimilate any food that comes its way, the Japanese giving it the most crude material. It needs little winter protection.

The Oriental poppy is a perennial worth waiting for. Seed must be sown where it is to remain, and the next year its immense flame-coloured blossoms and finely cut foliage will astonish the novice. It is a fast colour, too, that neither sun or wind can change. Planted in front of the *Astilbe Japonica,* that has its white feathery sprays about the same time, it forms a fine contrast, but its red over-powers everything when mixed inharmoniously with some colours. The front of the border will need little groups of plants, and the Montbretias, that are wintered as easily as gladioli, form a gleam of colour on slender talks, while the cardinal flower (lobelia) keeps up the brightness, and a clump of late dwarf phlox gives contrast. With plenty of room, a group of the perennial phlox in the wonderful new shades is a revelation.

Hollyhocks bloom the second year and make a good background. They are biennials, but last more than two years in well-drained soil. "Chaters" are double and of delicate colouring.

Larkspurs are both annual and perennial. The latter can be purchased, like many of these mentioned, as roots, without waiting for seedlings. Their clouds of pale or dark blue are effective in the border, and their lasting qualities are good if seed is kept from forming.

Scarlet Lychnis is one part and the deep cups of the blue and white Platycodon form another group, while Monarda (Horse Mint) will rear its showy heads, but too coarse for beauty.

The Day Lily (*Funkia subcordata*) blooms from August till October, fragrant and pure, its long tubes being the latest sweets to the belated wandering bee. The leaves are ornamental as a border plant with pale green heart-shaped leaves. Another variety has lavender flowers, while the leaves of another are variegated and margined with white.

Other late-flowering plants are Tritonias, that must be wintered indoors, and the fall Japanese anemone, that is not always hardy.

The darling of the ladies who are partial to yellow is the "Golden

Painting by George Ackermann of the Holman garden, Prince Edward
Island. Courtesy Prince Edward Island Museum and Heritage
Foundation.

Glow" (*Rudbeckia*). It has spread itself like an epidemic over the
country towns and byways, and is sturdy and faithful when flowers
are wanted for hardiness, and careless culture. If cut off when first
flowers are over, a new crop will come from the base, dwarf, but
pleasing because so colourful.

Hardy chrysanthemums are coming into favour, from the old
Belle Marguerite and Artemisias to the newer sorts that last into
November.

The hardy asters, or starwort, have been wonderfully improved
and make a cheerful group, proving the merits of civilization and
culture. . . .

Succession of bloom is necessary, for if all the varieties planted
blossomed in May or June, the border would be unattractive for
the rest of the season. For a short border, a crop of parrot tulips
will give gay flowers early, while dwarf white phlox could be set
at intervals between blue larkspur of the same height, and scarlet

salvia for the intervening plant. The latter is tender, and must be kept indoors in winter, but grows rapidly if cuttings are started in spring.

The foxglove is indispensable for an old-fashioned garden; its clean foliage and spikes of gay bloom make an effective appearance, but it is a little tender, and needs to be well covered up with leaves.

Taken altogether, there is more pleasure and less trouble in the perennial border than in any other division of the flower garden.

Mrs. Jack's joy with the perennial border is illustrated by a painting of the Holman Homestead, on Prince Edward Island, by George Ackermann. Careful scrutiny will reveal everything we have been considering. A handsome perennial border surrounds a lawn centered by a pseudo-fountain planting, surrounded in its turn by curved beds of annuals (note the ready greenhouse in the background). A vase, now planted to the brim, stands on the lawn, in its own flower ring. Four of the fashionable Lombardy poplars stand in an erect group. Vines grow against the ell wall. Two clearly defined styles of fence appear in the picture. The mixed planting of the border is carefully depicted. And, as always, just as in the atlas drawings, we have a happy horse and dog.

We can be grateful even today to these purposeful settlers for their statements, in still standing houses and still flourishing gardens, of the best in Victorian domestic landscaping transported across the Atlantic.

Specialty Gardens

OWARD the end of the nineteenth century, the world of American garden design began to develop unique departures from any plans as yet appearing in the rich library of plant and garden books. By this time, prosperity was apparently secured in the New World, ladies were beginning to come into their own as initiators and innovators, domestic and garden help was available for the training, and nurserymen were able to supply plants on demand.

The proliferation of what we can call specialized gardening—gardening in new designs, with new names and fresh purposes—may have sprung in large part from the impulses of women wishing to be free to garden on their own, for their own pleasure and within their own powers of maintenance. Earlier in the century, we saw ladies coaxed into light gardening, encouraged to study botany and to consider flowers as morally and socially significant. Many of the traditional trends in landscape gardening, both European and American, were male-inspired and male-oriented. Although it would be a long time before the American ladies' gardening image would rise, or sink, to advertisements showing them lifting heavy loads on balanced carts, or riding on mechanized lawnmowers, laughing the while, they began to command attention as a separate source of influence.

Ideas for special gardens, appearing like the medieval idea of separate rooms, not connecting with each other or continuing the one from the other but kept in enclosed units—walled, or hedged, or somehow secluded—may seem slightly rebellious rather than simply inspired by a love of flowers, but we can remember that the owners may have been truly bent upon escape as well as self-indulgence. The idea of a garden all of one color could pose problems for everyone, even the flowers. For instance, though famous ladies like Edith Wharton were known to have one-color gardens, and though artistic male garden designers in England fancied red borders and yellow "garden rooms," the idea of a one-color garden needs a voice to go with it. Especially when the color is blue. Because blue is rare in flowers generally, a challenge such as this to hybridizers resulted in excessive efforts to achieve blue in plants that had never been that color before, like tall border phlox (*Phlox paniculata*) and roses. Near successes resulted in pale purple phlox that looked bluish when not in the sun and in steel-gray rose petals that had to be called by metallic names and considered as made of pewter or silver to be bearable. Remembering the black tulip of history and literature, we can regard a black pansy with wonder, if not longing. But these one-color extravaganzas are so literally contrived that, without the owner's voice delightedly explaining all and the novice gardener's respectful attention, they lack universal appeal.

Similarly, secret gardens and scented gardens need accompanying sound tracks for visitors, as do the slowly evolving rock gardens where piles of gigantic boulders supporting Cupids gave way to simulations of rocky mountain slopes with appropriate plantings of alpine flowers. The wonder is, with all these individually oriented departures from the ordinary on the part of lady gardeners, that the idea of herb gardens so thoroughly and instantly caught on.

Although the heyday of herb gardening belongs in the twentieth century, the idea of incorporating these useful plants into vegetable and flower gardens had been around for centuries before enterprising late-nineteenth-century ladies seized upon it for an artistic and rewarding tour de force in ornamental gardens. Medieval monks had incorporated culinary and medicinal herbs into their garden plots. Medieval ladies had sat upon herb-planted banks while they cooled their feet in tiny fountains in walled gardens like outdoor castle rooms. Eminent writers on gardens, like John Parkinson, had allowed illustrations of

Mrs. Lawrence's rockwork garden, from *The Villa Gardener,* by J. C. Loudon, 1850.

especially hairy herbs like borage into his pleasure gardens, because the ladies enjoyed using the extra stitches to put them into their fancy work. In the Appendix we list the herbs recommended by M'Mahon from his early-nineteenth-century nursery in Philadelphia and by another Philadelphia nurseryman, Thomas Bridgeman, in midcentury; but we must emphasize here that these were all included in the vegetable gardens.

Until the end of the nineteenth century, no one had conceived of a so-called herb garden for pure pleasure. At that point, the tasty and healthful herbs were drawn out of association with the vegetables and became the crowning features of intricately designed small gardens, like the medieval knot gardens, where strands of clipped herbs crisscrossed each other, with all space between filled. Whereas gray and green ropes of trimmed foliage passing over and under each other were emphasized in the medieval plots, these new gardens were filled with all sorts of scented, flavorful, and useful little plants.

Even the word *herb* began to have a special meaning. *Herb*—the early English word, indeed, for *plant*—began to refer only to what ladies were willing to place within their herb gardens.[1] An interest in French provincial cookery has been said to have promoted the pretty little formal creations in green and gray, but more likely it was the other way around. In any case, the herb garden, as such, arrived in all its charm.

Another specialty type of gardening beloved by ladies, which had been used in moderation for years, brought roses out of herbaceous borders and into borders of their own—and also over arbors, up veranda posts, and in formal single beds, usually in the middle of a lawn. As Italian gardens became popular, rose gardens appeared where geometrically arranged beds were surrounded by pergolas, trellises, bowers, and posts with chains strung between them; roses were trained even on the chains.

With the growing interest in specialty gardens, one could have a bog garden, with gigantic leaves swooping down from higher banks to slip into the murky waters of a lily pool—a garden effective when looked down upon but not to be wandered in. In contrast to the other small gardens considered here, which were the province of women, bog gardens belonged almost solely to adventurous men in hip boots.

Wild gardens became a vogue in England (under the aegis of the great William Robinson, in his war against ribbon bedding, and of Gertrude Jekyll, in her battle for harmonious colors), but only the most advanced elderly New England aunts seem to have taken to them in America during the late nineteenth century, perhaps because the wilderness seemed too close to invite for a social call.

With specialty gardens of reasonable size, capable of being fussed over and of showing a response to the care lavished upon them, ladies were not dependent upon professional designers for their layouts, or upon books for information on maintenance. They could concoct for themselves and often did. Downing would have been surprised—and pleased.

The wonder, in all the newly burgeoning world of totally new influences in garden design, is that so many were passed over or by. Japanese gardens, for instance, began to appear, although not at the pace they acquired in the next century, when they were so falsely promoted

遠樹無枝
山腰雲塞
樓閣樹塞

Japanese landscape lesson illustrating techniques for depicting different sorts of trees. From an early eighteenth-century how-to-do-it manual. Courtesy Ipswich Historical Society.

as easy to maintain. A touch of the Orient in a railing or a curved roof, a great concentration of wistaria and irises, with perhaps a bronze crane of more than lifesize proportions, evoked yet another garden style. Only these superficial aspects of the true Japanese garden drew the attention of American gardeners in the nineteenth century, although the naval surgeon who sailed with Commodore Perry brought first notice of those gardens where a world is created in miniature, so beguilingly that the observer feels reduced to a size able to enjoy the landscape in scale. Dr. Joseph Wilson's letter to Caleb Cope in Philadelphia was "ordered spread upon the minutes" of the meeting of the Pennsylvania Horticultural Society in February 1855, so the news spread.[2]

U.S. Ship Supply, Jany. 26–1855

CALEB COPE, ESQ.
   My Dear Sir—

   The leading character of Japanese horticulture is its attempt to represent rural landscapes on a small scale by means of rock works, small pools, flowers and trees. I think the most exquisite specimen I have seen in this style was at Napa—Lewchew. The whole enclosure is about ten feet by twenty-five . . . feet bounded on one side by the house and the other three sides by board fence. This contained three pools at different levels and slightly overhanging each other, full of fish—about thirty square feet of surface in the aggregate—some of the fish were as much as fifteen inches in length, although they were generally less than half that length. These pools are made of sandstone of irregular shape— merely by cutting cavities in the upper side. They are placed one above the other in such a way as to form one side of the miniature mountain. There is a good path to the top of this mountain, with a stone seat and a convenient place to empty a bucket of water about once in twenty-four hours, and the arrangement is such that in draining, it waters all the plants and supplies all the pools, dripping regularly from one pool to the other to the lowest, whence an underground channel finally discharges the surplus. There are about six kinds of trees on this mountain—generally the pines, etc. of the adjoining country, exceedingly dwarfed, though quite too large for the mountain. There are venerable looking old pines about ten inches high—though some are as high as two and one-half feet. This extreme dwarfing is effected by planting a seed or a very young plant in a little very good soil in small excavations in the rock—where it never wants for its regular supply of water drawn from the fish pools—and by trimming off some of the branches and distorting others to the forms which most nearly resemble the old tree. This mountain has likewise some very pretty green grass on the slopes—a rustic bridge,—and quite a number of cottages. The first view of this miniature landscape is rather pleasant—especially as the whole is very neatly kept . . . but it is not so easy to admire it when you reflect that it is the most effectual arrangement which could be contrived for breeding mosquitoes. The same style prevails . . . doubtless in Japan generally. The meanest houses generally have at least one dwarf pine and a pool a yard in diameter for the accommodation of a few bright-colored fish.

JOS. WILSON, JR., PA Surgeon,
U.S. Navy

Chinese gardens, identified chiefly by the ornamental pots with which all families connected to the Oriental trade were familiar, paled in concept beside these miniature Japanese worlds. Of the Japanese garden fantasy, only minor features—like the dwarfing of forest trees—have become modern specialties. But the creation of these landscapes may have seemed more of an undertaking than the lady gardeners could handle.

In the very late nineteenth century, a gardening phenomenon arose whose rise and fall occurred suddenly but proved worth recording. It can only be called *horizon gardening*. Seen before as a province of French monarchs, it arrived with the sudden emergence of great fortunes. It would not be worth noticing except for its benefits, which have proved of lasting value to the general public and the cause of conservation.

Each early settler had carried in his heart an image from his origins, to be adapted for his comfort and reassurance on the new land. Log cabins came with Swedish woodsmen, cottage gardens with English villagers, patios with Spanish missionaries, town courtyard gardens with French traders at rivers' mouths. With financial success, different

Chinese garden in pots.

styles developed, some even from the books written to "improve" neighborhoods. Sensible reference to available materials and the general social standing of the individual in the community was considered good taste. Although a certain amount of competition might show up on communicating front lawns—in the stances of iron deer and dogs or even in the comparative brightness and elaboration of fancy set pieces in flower beds—the general idea was a seemly adaptation to the existing norm.

A new tendency toward competitive ostentation began to creep into the American scene. The Chicago World's Fair was conceived as an opportunity to show the world that Americans were no longer dependent upon the Old World for either arts or manufacturing skills. The Great White City was graced with huge buildings, classically inspired although carrying modern business names, which had been competitively executed by American architects. Even domestic architecture, as ordered by the very wealthy, took on the flavor of ancient grand buildings. Marble palaces redesigned as summer residences were joined by reproductions of French châteaux and Norman keeps. Surrounded by lavish representations of famous gardens, these edifices are referred to today as belonging to a "golden age" of American architecture and landscape gardening. It seemed as if the public's developing critical faculties had been numbed, to be replaced by a demonstration of what money could buy, under the sponsorship of a new society of the very rich.

Strange—when the early part of the century had been so philosophically and proudly devoted to the establishment of original American norms—to find this sudden copying of European grandeur. Stranger, also, to find that satisfaction suddenly required the acquisition of vast spaces. Although today these imitation European gardens are seen chiefly in the ruins of pergolas made of cement instead of marble, in great terraces that lead to tangled woods instead of mazes, in balustrades of wood rather than stone, and in water gardens returned to bogs, the idea of ownership-to-the-horizon has resulted in great benefit to the general public.

A previous insistence on a horizon-based boundary may have been implicit in Le Notre's feeling that the end of each grand allée and central axis should appear to be the sky.[3] But nothing in the developing of American garden design had foretold the insistence of leading American landholders upon owning all the land that met their eyes.

In one case belonging to the early twentieth century in execution but stemming from the late nineteenth-century rise of great fortunes, the owner faced the Atlantic Ocean and so needed to acquire only the 180 degrees of field and marsh, creeks and islands, to his rear. One of the views from the replica of a great English country house included an enormous beach, which had been too distant for public enjoyment in the years preceding mass transportation. Today the public has permanent access to an unspoiled beach in a world prone to commercialize its shorelines.

An even grander concept of a summer home overlooked a vast lake in Vermont. The rear view to the horizon included farms and fields, orchards, barns and pastures, fences and walls and roads. The style of the house, settled after many ambitious architects were consulted, is Burgundian, with patterned slate roofs and cylindrical turrets. The stables, built about a courtyard and also Burgundian, had an entrance door large enough to admit a coach-and-four. The barns, also built about a courtyard and in the same style, housed twenty work teams and their drivers. Today, the public roams the vast expanses and listens to concerts in the enormous rooms.

The acquisition of the land involved had become a family legend. The first owner and builder determined to have "an English park," which he defined as "the shadows of great trees on close-mown turf." Calling his men to a conference on a Saturday morning, he had handed each one an envelope with a farmer's name and a sum of money. The next morning after church, one of the farmers announced apologetically to his peers that he had decided to sell his farm. So, it appeared, had they all.

These places are well known and well used today. We do not identify them here only because first acquaintance was when they were privately owned and family stories have their rights.

The greatest of any of these to-the-horizon designs has to do with the esteemed Frederick Law Olmsted and is worth considering both in itself and for that alone. In the Blue Ridge Mountains of North Carolina, near Asheville, stands Biltmore, which has been called the world's largest single-family house. Built to resemble the Blois château in the Loire valley, this version of the French Renaissance style has well over 200 rooms. The site was discovered in 1886, by the son of the richest man in America at that time. He gradually secured rights to more and more land, eventually acquiring over 100,000 acres. Design of the house was undertaken by the well-known architect Richard Hunt. The sixty-four-year-old Olmsted was persuaded to lay out the grounds. Both house and grounds were to be finished as soon as possible.

This was to be the last large commission Olmsted accomplished, and into it he put the accumulation of his experience. All the inescapable demands of the times and its fortunes had to be met, though on a scale larger than had ever been attempted; for example, the English walled garden must be larger than any made before in the United States. Also included were an enormous Italian garden of great pools upon a vast terrace that stretched away from the house and a gigantic lawn where a reflecting pool could catch the towering entrance of the main facade, with its circular ramp staircase like the one at Blois. Conservatories and greenhouses covered four acres. On one side of the main entrance was a *rampe douce,* copied from the one at Vaux-le-Vicomte, where Le Notre carried out his first important commission. Trees planted in quincunxes around the lawn were bordered by narrow bands of water, as at another French château.

To Olmsted, this last commission must have appeared to be both a valuable opportunity and an exercise in negation of what he had worked for. He had spent his life in consideration of the well-being of the general public (particularly in its access to fresh air and beauty), had begun by loving untroubled natural beauty, and had never ceased to try to protect or to imitate it for the thousands of city dwellers to whom he had given a reprieve from misery and overcrowding. Although in this vast estate much effort and money must have been expended (at which he could only wonder privately), once the conventional trappings were taken care of, he was able to devote himself to his old and early loves: a vast shrub garden, an azalea slope, a pinetum, an arboretum, and more.

The entrance drive was made to curve slowly for miles. To hear about it is to remember the drives of the past, like Parmentier's to Dr. Hosack's, Downing's to Newburgh, and those of Vaux and Olmsted in Central Park. From the lodge gates, the road first winds beside a stream and then rises past a grove of oaks to an open field of corn, on beside a beech grove and a pasture. A ravine through hemlocks and pines reveals a brook and a specially designed waterfall. The road continues to rise through a forest full of native dogwoods. Everywhere are specimen trees. In line with another of Olmsted's standby features, there is a "ramble," obviously of considerable extent.

Olmsted's contoured map of 1890 includes 7,000 acres. He must have felt joy in commanding a forty-acre nursery of young trees and in importing the loveliest trees and shrubs from around the world. He must have taken particular pleasure in basing some of his plans upon memories of his first trip abroad when, as a young man, he undertook a walking tour with his brother and a friend and wrote *An American Farmer in England*. Biltmore has echoes of Birkenhead Park, near Liverpool, the first great park Olmsted had ever seen and the one that had enchanted him with its intention of furnishing a classless pleasure ground, where one sort of landscape use progresses to another, through bridges and open fields, groves and winding paths and roads. Olmsted uses touches of Chatsworth, too, at Biltmore, perhaps comforted by the thought of its vastness.

But in the end Olmsted saw at Biltmore an opportunity to exhibit the beauty and the use of an entire landscape given over to forestry—all the way to the horizon. He had enlisted the advice of Charles Sargent, authority on trees of the Arnold Arboretum, and of Gifford

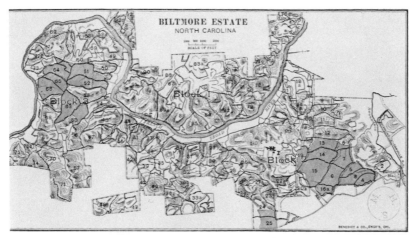

Biltmore Estate, North Carolina.

Pinchot, first American-trained forester. He foresaw a great American experiment in growing properly a vast forest of both native and imported trees—an arboretum that (in keeping with all at Biltmore) would be the largest in the world.

Olmsted's health and his client's fortunes declined. He left Biltmore in 1895, and in the end 80,000 acres were sold to the newly formed United States Forestry Service to become the Pisgah National Forest.

Which is, again, a happy outcome.

NOTES TO CHAPTER 20

1   The *h* was not sounded in America, although it had been sounded in England, according to the O.E.D., since 1730. So one said "an erb" and not "a herb" when the ladies became really organized.

2   Wilson describes exactly the sort of garden I encountered in Japan seventy-five years later and have never been able to defend successfully against the wastes of pebbles, moss, and stones now firmly identified as "standard" Japanese gardening; thus, I am particularly happy to include it here.

3   In Williamsburg today, the rising of a great slope of turf (which hides the railway at the bottom) in the Governor's Garden allows this garden to seem to reach into the sky, a brilliant solution.

Afterword

I s it a coincidence that many leading visionaries who were able to change the American view of the environment—Marsh, Downing, Olmsted, Eliot—were boys who wandered their countrysides? As young men, they had all suffered disabilities that kept them from prolonged schooling and the rigors of commercial or professional careers. As adults, they became of greatest value to their country by realizing the dreams of their youth. To them we owe the American landscapes, parks, gardens, forests, and national preserves.

We have seen Downing's reverence for a good American elm and Marsh's for an oak. We have also seen Olmsted's involvement with trees and all their uses, including his novel idea for preserving a forest of enormous extent by so managing it as to make it pay for itself and its own costs of renewal. A little book published especially for the Chicago World's Fair in 1893 by Gifford Pinchot, consulting forester for Biltmore Forest, showed how far and fast they had come.

Let Downing bring us back to our beginning, with the view from the summerhouse. In the 1849 edition of his *Landscape Gardening,* Downing mentioned that in

the environs of New Bedford are many beautiful residences. Among these we desire particularly to notice the residence of

296

"View in the Grounds of James Arnold, Esq., New Bedford."
From Walter Whitehill's papers for the centennial celebration of the
Arnold Arboretum.

James Arnold, Esq. There is scarcely a small place in New England where *pleasure-grounds* are so full of variety, and in such perfect order and keeping, as at this charming spot; and its winding walks, open bits of lawn, shrubs and plan groups on turf, shady bowers and rustic seats, all most agreeably combined, render this a very interesting and instructive suburban seat.

He includes a picture of the ideal summerhouse as a "View in the Grounds of James Arnold, Esq." It serves as a symbol of American nineteenth–century gardening because (in addition to its absorption, with much else of importance, into the Arnold Arboretum) it illustrates M'Mahon's summary of the good fortune of the American people, whom he saw as "so universally possessed of landed property" as to be nationally blessed "with comfort and affluence."

Appendix: Plants Most Commonly
Used in Nineteenth-Century
American Gardens

In order to make authentic restorations of those gardens created during the nineteenth century across the so rapidly evolving United States, it is important to offer a survey of the most popular available plant material. In my *Early American Gardens: "For Meate and Medicine,"* I listed the plants we know were available from contemporary, documentary evidence, using the familiar names first. I followed the same order in *American Gardens in the Eighteenth Century: "For Use or for Delight,"* with a comprehensive list drawn from letters, account books, and plants offered for sale by the earliest botanical collectors and nurserymen. With this book, *American Gardens of the Nineteenth Century: "For Comfort and Affluence,"* I have presented the lists of plant material with the botanical names first, followed by the familiar names where they were most used. The popularization of the Linnaean system and the discoveries of so many new plants, which enlivened the times and the gardens, resulted in plants with no familiar names—as for instance, *Dahlia*—whereas many familiar names made do for quite different plants—as snowberry, burning bush, trumpet vine, and so on. And again, rose of Sharon was *Hypericum* in England and *Hibiscus syriacus* in America. As the whole system of garden layout changed so radically, it seems reasonable here to group the plant material by the uses to which it was put. The Victorians expected a good deal from their plants, though not as much as the early settlers in the seventeenth century had expected from *every* plant. I have grouped vines, for instance, for decorating houses, trees for parks and lawns, shrubs for shrubberies, and so on, with a foreword for each group. Roses, which were beginning to be grown separately instead of as "border trees," I have listed by themselves. And I have included a listing under "Herbs," although it was a twentieth-century fancy to segregate ornamental plants from those used in cooking and other household emergencies.

And I have attempted a broad survey of the times by depending upon the books of advice written by a number of the most popular authorities at different dates throughout the century. Quotations are followed by the author's name. The names of other seedsmen who carried the plants follow his.

The botanical names have been updated by reference to L. H. Bailey, *Hortus III* (1976). A bibliography of the reference books used, with their dates of publication and the editions consulted, follows the Appendix.

I *Annuals*

The sudden advent of a host of "tropical" plants from South America and Mexico into North American gardens and into the gardens of eminent Victorians in England was responsible for the wave of formal bedding-out that took the mid-nineteenth-century garden world by storm and affected public gardens forever after. Tiny flowers of brilliant colors could be raised under glass while the beds they were destined to fill were already full of blooming bulbs. When the bulbs faded, the annuals could be popped into position to cover their dreary demise. This operation created and solved many problems for professional gardeners. And they were professional. A skill with seeds and seedlings, a way with glass and stoves, a hand for nipping and pruning, an eye for the maintenance of the same color for many square feet of bloom—these practices were beyond all but the most ardent amateur. To make a floral clock, an uptilted map of some foreign country, a far-flung flag, a simulated fountain, or even just an extravaganza of leaves, spotted and striped, took special skill and patience.

A. J. Downing's dislike for any bare ground showing in flower borders and beds was, as he admitted, impossible to appease without the use of annuals planted en masse.

Abronia   Sand Verbena
> *A. umbellata.* "A beautiful annual . . . long trailing stems . . . elegant flowers . . . delicate lilac . . . white centre . . . deliciously fragrant" (Breck).

Adonis annua   Pheasant's Eye
> *A. aestivalis.* Flos adonis, or tall bird's-eye (M'Mahon).
> *A. autumnalis.* Pheasant's-eye. "Foliage handsome, flowers blood-red, hardy" (Breck, M'Mahon, Buist).

Ageratum   Floss Flower

     A. mexicanum (now *A. houstonianum*). "Light blue, very pretty, but not remarkable for beauty" (Breck, Bridgeman, Buist).

Agrostemma   Corn Cockle (see also *Lychnis,* under "Perennials")

     A. coeli-rosa. Smooth-leaved campion (now *Lychnis coeli-rosa,* rose of heaven) (M'Mahon).

     A. githago. Corn rose campion (now *Lychnis githago,* corn cockle) (M'Mahon).

Alyssum   Madwort

     A. maritimum. Sweet Alyssum. "Fine effect when planted in masses" (Breck, Buist).

Amaranthus   Amaranth

     A. caudatus. Love-lies-bleeding. "At a little distance pendant spikes look like streams of blood. It is sometimes called in France *Discipline des religieuses*—the Nun's Whipping-rope" (Breck). "Love Lies Bleeding, red and yellow variety" (Buist, M'Mahon).

     A. hypochondriacus. Prince's feather. "Numerous heads of purplish-crimson flowers, suitable for shrubbery" (Breck, M'Mahon).

     A. tricolor. Joseph's coat. "Chief beauty . . . consists in its variegated leaves" (Breck). Miller knew no handsomer plant "in its full lustre." Gerard called it Floramor and compared it to a parrot's feathers: "It farre exceedeth my skill to describe the beauty and excellencies of this rare plant." "Three coloured Amaranth" (Bridgeman).

Anagallis   Pimpernel

     A. arvensis. "Pimpernel or Poor Man's Weather-glass. Red or Blue" (Breck, M'Mahon).

Antirrhinum   Snapdragon

     A. elegantine (now *Linaria vulgaris*). Butter-and-eggs; fluellin, or toadflax.

     A. majus. "Perennial or biennial around Mediterranean. Treated here as an annual. The great purple snapdragon, described by Gerard" (Breck).

     A. viscosum (now A. *asarina*). Clammy snapdragon (M'Mahon).

Argemone   Prickly Poppy, Argemony

     A. grandiflora. "Large flowers, delicate white petals, splendid appearance, superior to others" (Breck).

     A. mexicana. Prickly poppy (M'Mahon). "*Argemone* or Prickly Poppy— Yellow, cream-colored and white" (Bridgeman).

Asclepias   Milkweed, Silkweed

     A. curassavica. "Swallow wort, orange and red flowered" (Buist).

     A. cornuti (formerly *A. syriaca*). "Parkinson's Virginian silk" (Breck).

     A. tuberosa. Butterfly weed (now considered perennial). Pleurisy root (Breck).

Aster   China Aster (see *Callistephus*)

Avena   Animated Oats

   A. sterilis. "Sometimes grown as an object of curiosity because of the hygrometrical properties. After seeds have fallen their strong beard is so sensible to alternation in the atmosphere as to keep them in motion . . . resembling some grotesque insect crawling upon the ground" (Breck).

Balsamina (see *Impatiens*)

Bartonia

   B. aurea (now *Mentzelia lindleyi*). "Golden-flowered bartonia. Native of California. Profusion of showy flowers of a fine golden-yellow color" (Breck, Bridgeman).

Bladder Ketmia (see *Hibiscus trionum*)

Bluebottle (see *Centauria cyanus*)

Brachycome   Swan River Daisy

   B. iberidifolia. "White, pink or blue . . . grown in masses or beds; foliage delicate" (Breck).

Briza

   B. maxima. Quaking grass. "Sometimes cultivated as a border flower; the spikes of the grass are elegant, when dried, and help to make up a bouquet of immortal flowers" (Breck).

Browallia   Bush Violet

   B. elata (now *B. americana*). Blue amethyst. "Native of Peru. Bears an abundance of small bright-blue flowers. Variety with white flowers" (Breck).

Cacalia (see *Emilia*)

Calandrinia   Rock Purslane

   C. ciliata. Rose-colored calandrin. Red maids (Buist, Bridgeman).

   C. grandiflora. "Foliage is purple on the underside" (Breck).

Calendula

   C. officinalis. Pot marigold. "Common to the gardens time out of mind and formerly much used in soups and broths" (Breck, M'Mahon).

Calliopsis (see *Coreopsis*)

Callirrhoe   Poppy Mallow

   C. pedata. Introduced from Texas. Crimson mallow-shaped flowers. "Handsome" (Breck).

Callistephus   China Aster

   C. chinensis. "Double China Aster now known as German Aster, Full-quilled, dahlia-sized, various habits . . . Possess exceeding beauty" (Breck). French called it "Reine Marguerite" (Breck). "China Aster or Queen Margareta, in great variety. The last imported Asters are of extraordinary beauty" (Buist).

C. *chinenis* vars. *alba, rubra, striata, purpurea,* etc. "Aster Chinese and German, white, red-striped, purple, etc." (Bridgeman).

Candytuft (see *Iberis*)

Canna Indian Shot

C. *indica.* "Introduced into England in the sixteenth century but only this one variety. In the early nineteenth century a great number were found in South America" (Breck).

Cardamine Cuckoo Flowers, Lady's Smock

C. *pratensis.* "Double cardamine only deserving a place in borders" (Breck).

Catchfly (see *Silene*)

Celosia Cockscomb

C. *cristata* and C. *lutea.* "Cockscomb, crimson and yellow. Common in most gardens" (Bridgeman, Buist).

Centaurea

C. *cyanus.* Bluebottle. Bachelor's button.

C. *americana.* "American Centaurea—Basket Flower." "Discovered by Nuttall" (Breck).

Chrysanthemum

C. *carinatum.* "Tricolored chrysanthemum—Native of Barbary. Disk of the flower purplish brown, inner circle of the rays yellow, margined with white; very pretty" (Breck).

C. *coronarium.* Crown daisy. "Garden chrysanthemum—One variety with white and another with yellow flowers. Double varieties alone are worth cultivation" (Breck).

C. *parthenium.* Feverfew (now *Matricaria capensis*). "Double—worthy of border planting" (Breck).

Clarkia (named for Clark by Pursh)

C. *elegans.* Elegant clarkia. "Rose colored. Found in California by Mr. Douglas" (Buist, Breck, Bridgeman).

C. *pulchella.* "Beautiful Clarkia—Native of the Rocky Mountains. Discovered by Captain Clark in his expedition with Captain Lewis, to the Columbia River. Light purple flowers. Variety with white flowers" (Breck).

Cleome Spider Plant (now C. *unguiculata*)

C. *grandiflora.* "Pale pink-purple. Beautiful in the garden, but withers very quickly when cut" (Breck). "Lilac flowered Spider Plant" (Buist).

C. *pentaphylla* (now C. *gynandra*). "Flowers pure white. Odor . . . most offensive" (Breck).

C. *spinosa.* Giant spider plant. "Spider flower—Spike of beautiful white flowers" (Breck, Bridgeman).

Clintonia

    C. elegans (now *Downingia elegans*). Elegant clintonia. "Rich blue flowers in great profusion" (Buist, Bridgeman).

    C. pulchella (now *D. pulchella*). Pretty clintonia. "Native of California. Blue with broad white spot at the centre, stained with rich yellow" (Breck).

Cockscomb (see *Celosia*)

Coix lachryma-jobi Job's Tears

    "Job's Tears—Tropical grass from the East Indies. Appearance of its shiny, pearly fruit . . . when suspended on its slender pedicels, resembles in no inconsiderable degree a falling tear. The flowers are destitute of beauty" (Breck).

Coleus Flame Nettle, Painted Leaves

    C. verschaffeltii (now *C. blumei*). "Popularized in mid-century by the German 'florist' although other varieties were known earlier in the century. Grown only for its leaf and generally from cuttings for mass-effects . . . Var. *aurea zebrina* has yellow stripes" (Breck).

Collinsia

    C. bicolor. "White and light purple. Beautiful" (Breck, Bridgeman). "Good in masses" (Buist).

    C. grandiflora. "Blue and purple" (Breck).

Collomia c. coccinea. "Bright carmen red. Desirable dwarf annual" (Breck).

Commelina Day Flower

    C. coelestis. Blue spiderwort. "Sky-blue Commelina—Blue flowers . . . cannot be excelled, and its profusion of blossoms renders it deserving of cultivation in every flower garden" (Breck). "Plant a circular bed, eight feet across with a four-foot circle of Commelina at center and a surround of spring-sown plants, all six inches apart. The Commelina roots will live if taken up and thus used the blue cone effect at the center of the bed is achieved" (Breck, Bridgeman, Buist).

Consolida ambigua Rocket Larkspur (see *Delphinium* for comments)

Convolvulus

    C. minor (now *C. tricolor*). Dwarf morning glory. "A bed of same planted two feet apart will interlock and be a handsome sight all summer" (Breck, M'Mahon).

Coreopsis (interchangeable with *Calliopsis*)

    C. atrosanguinea. Crimson (Buist).

    C. coronaria. Pale yellow (Bridgeman).

    C. drummondii. "A fine bedding plant" (Breck). "Discovered by Mr. Drummond and named after him" (Bridgeman, Buist).

    C. tinctoria. "Introduced by Nuttall who found it in Missouri. Called 'Nuttall's weed.' Beautiful when confined. From this many superb varieties have been obtained" (Breck).

   C. *tinctoria* var. *atropurpurea*. "Dyeing Coreopsis. Dark-flowered variety. . . . Deserves a place in every flower garden. Splendid in masses. . . . Flowers have yellow centers, surrounded by purple circle, edged with a fine scarlet" (Breck).

Crepis

   C. *barbata*. "Purple-eyed Crepis—Uncommonly beautiful sown in masses and the plants thinned out to 18". Rays of light yellow, centres purple-brown" (Breck).

Crucianella   Crosswort

   C. *stylosa*. Recommended by Mead to lady gardeners.

Cuphea   "Mexican—both scarlet and variegated" (Bridgeman).

   C. *llavea*. Red, white, and blue flower.

   C. *procumbens*. Recommended by Mead to lady gardeners.

Datura   Thorn Apple

   D. *ceratocaula*. Horn-stalked datura. "Highly ornamental" (Breck).

   D. *meteloides*. "Very splendid" (Breck).

   D. *quercifolia*. "White, the nerves of a fine pink, shaded with a dull purple . . . leaf somewhat like the oak. We can truly recommend this as an ornament in the garden" (Breck).

   D. *stramonium*. Jimson Weed. "Light purple or white. Poisonous" (Breck).

Delphinium   Larkspur

   D. *ajacis*. "Dwarf Rocket Larkspur—Greatest summer ornament of the garden" (Breck).

   D. *consolida* (now *Consolida ambigua*). "Branching Larkspur—Flower blue, white, pink and variegated." "A bed of the double varieties . . . is equal in beauty to a bed of fine hyacinths" (Breck).

Dianthus

   D. *barbatus*. "Sweet William—A bed of fine sorts presents a fine sight" (Breck).

   D. *chinensis* (annual, biennial, and short-lived perennial). "Indian pink—Dwarf habit. Without fragrance. Crimson and dark shades of that color approaching to black, are often combined with the same flower, with edgings of white, pink, or other colors" (Breck). "China Pink of every shade" (Bridgeman). "Many fine double varieties" (Buist).

Didiscus (see *Trachymene*)

Dracocephalum   Dragonhead

   D. *Moldavica*. Moldavian balm (Breck, M'Mahon).

Echeveria   Hen and Chickens
"Valuable for 'flat bedding'" (Breck).
Emilia   Tassel Flowers, Flora's-Paintbrush
   E. javanica (formerly *Cacalia coccinea*). "Scarlet Tassel. Flower neat of easy culture" (Breck). "Venus Paintbrush" (Buist).
Eschscholzia   California Poppy
   E. californica. "California Poppy—Brilliant shining yellow. Scarcely any plant produces a greater degree of splendor than this" (Breck, Buist).
   E. crocea. "Saffron-colored California Poppy—Dark, bright saffron color. . . . Most suitable for producing distant effect, plant in masses, needs staking" (Breck, Buist).
Euphorbia variegata (now *E. marginata*)   Snow-on-the-Mountain
   "Native of Missouri and Arkansas Territory. A most elegant species. Highly esteemed" (Breck, Buist, Bridgeman).
Evening Primrose (see *Oenothera*)

Fumaria
   F. fungosa (now *Adlumia fungosa*). Spongy-flowered fumitory. Mountain fringe, Allegheny vine (see also *Adlumia-fungosa,* under "Vines").
   F. glauca (now *Corydalis glauca*). Glaucous fumitory (Breck, M'Mahon).

Gaillardia   Blanket Flower
   Annuals derived from *G. pulchella,* perennials from *G. aristata.*
   G. bicolor. Two-colored gaillardia. "Large crimson disk, surrounded by a ray of fine yellow . . . showy appearance . . . deserving place in every garden. . . . Too tender for our winter" (Breck).
   G. picta. Painted gaillardia. Perennial. "Handsome plant, crimson and yellow flowers"; variety of *G. pulchella* (Breck).
   G. pulchella and *G. drummondii* are synonymous. "Natives of Mexico" (Breck). Orange and crimson (Buist, Bridgeman).
Gaura
   G. lindheimeri. "From Texas. One of the finest we have received. Fine for bouquets, may be cultivated as an annual" (Breck).
Geranium (see *Pelargonium*)
Gilia
   G. capitata. "Headed Gilia—Clusters of blue or white flowers. Pretty" (Breck, Bridgeman). "Scarcely anything can be prettier than this plant when thickly filling a bed a few feet in length and breadth" (Breck).
   G. tenuiflora (now *Linanthus liniflorus*). "Slender flowered—New from California. Pale rose, slightly streaked with red outside, and a fine violet inside" (Breck).

G. *tricolor.* Bird's-eyes. "Three-colored Gilia—Native of California. Yellow eye surrounded by a purple ring, bordered by pale blue or white. Something more showy is wanted to suit the common taste" (Breck).

Godetia (see also *Clarkia*)

G. *densiflora.* "Purple flowers in clusters" (Breck).

G. *lyndleyana.* "Lindley's Godetia—Either white or bluish with a rich purple blotch on each petal which gives great beauty to the flowers" (Breck).

G. *rubicunda* (now *Clarkia rubicunda*). "Ruddy Godetia—Native of California. One of the best annuals" (Breck, Bridgeman). "Other varieties worthy of cultivation are G. *lepida, quadrivulnera, purpurea, bifrons,* etc." (Breck).

Gomphrena   Globe Amaranth

G. *globosa.* "Globe Amaranth—White, purple and striped are desirable. A popular *immortelle*" (Breck, Bridgeman).

Helianthus   Sunflower, Girasole

H. *annuus.* "Garden Sun-flowers.—Too well known to need description. Great exhausters of the soil. Double dwarf sunflower is more desirable" (Breck).

Helichrysum   Everlasting Immortelle

H. *bracteatum.* Strawflower. "Golden Eternal Flower—Variety with white flowers" (Breck).

H. *macanthrum* (now *H. monstrosum*). "Large Everlasting Flower—Variety with white tipped with red; another yellow tipped in the same way. . . . Admired on account of the beauty of their flowers when dried. Highly prized for winter mantel bouquets and ornaments for vases, etc." (Breck).

Heliotropum   Heliotrope, Cherry Pie

H. *peruvianum* (now *H. aborescens*). "Chiefly admired for its fragrance" (Breck).

Hibiscus trionum   Flower-of-an-Hour, Bladder Ketmia (Bridgeman)

Iberis

Native of Iberia, now Spain. "Candytuft" now applied to all species.

I. *amara.* White Candytuft. Rocket. "Hardy. No little beauty" (Breck).

I. *coronaria.* Rocket Candytuft. "Pure white racemes resemble Double White Rocket at a distance" (Breck).

I. *umbellata.* Purple Candytuft. "First discovered in Candia. The original Candy-tuft. Purple flowers" (Breck).

Ice Plant (see *Mesembryanthemum*)

Impatiens

I. *balsamina.* "Touch-me-not. Double Balsam is one of the most prominent ornaments of the Garden. Crimson, scarlet, light and dark purple, rose, blush white. . . . some are variegated, or spotted with various shades of red and purple which are decidedly the most elegant. Native of India, Malaya and China" (Breck). "Commonly called Balsamene—elegantly formed" (Fessenden).

Jacobaea (see *Senecio elegans*)
Job's Tears (see *Coix lachryma-jobi*)

Lantana camara   Shrub Verbena, Yellow Sage
   "For baskets" (Breck).
Lathyrus odoratus   Sweet Pea
   "Sweet pea—One of the most beautiful . . . most fragrant . . . most popular annuals which enrich the flower bed" (Breck).
   L. albus. White.
   L. caerulis. Blue sweet pea (M'Mahon).
   L. carneo. Old painted lady.
   L. roseo. New painted lady, or scarlet sweet pea.
Lavatera trimestris   Tree Mallow
   "Common Lavatera—Species resembles those of *Malva*. Handsome, white and pink" (Breck).
Limnanthes douglasii   Meadow-foam
   "Mr. Douglas' Limnanthes—Native of California. Numerous fragrant flowers, deep yellow, the extremities of the petals being white. In the native habitats, growing by sides of lakes, river, etc." (Breck).
Linaria vulgaris   Butter-and-eggs, toadflax
Linum   Flax
   L. grandiflorum. "Brilliant crimson flower" (Breck).
Lobelia erinus "Much used for edging" (Breck).
Lunaria purpurea   Honesty, Moonwort, Satin Flower (Bridgeman)
Lupinus
   L. cruckshanksii. Cruckshank's lupine. "Elegant white flowers shaded with yellow and blue or purple" (Breck).
   L. hartwegii. Hartweg's lupine. "One of the most beautiful. Spikes of rich blue" (Breck).
   L. varius. "Smaller foliage, fine blue flowers" (Breck).
   Old varieties cultivated are:
   L. albus. White.
   L. hirsutus. Great blue.

L. luteus. Fine yellow flowers (Breck, M'Mahon). "Required to be partially shaded" (Bridgeman).

L. texensis. Texas Bluebonnet.

Malope
 M. grandiflora. "Grand flowering Malope—Mallow tribe. Desirable plant for giving a distant effect" (Breck).
 M. trifida alba. "More delicate in habit" (Breck).

Martynia (now *Proboscidea*)   Unicorn Plant
 M. fragrans. "Flowers are large, resembling the Gloxinia . . . delicate rosy lilac, blotched and shaded with bright crimson" (Breck).
 M. proboscidea. "Common Martynia—Often cultivated in vegetable gardens for its capsules, which, when green and tender, make a fine pickle" (Breck). Cuckold's horn (Breck). Unicorn plant (Bridgeman).

Matthiola   (now *M. incana* var. *annua*)   Stock Gillyflowers, Ten-Week (or Intermediate) Stock
 "Many varieties . . . with fragrant, handsome flowers. German stocks are very much celebrated" (Breck). "If I were to choose amongst all biennials and annuals, I should certainly choose the stock. Elegant, beautiful, showy, and most fragrant flower" (Cobbett).

Mentzelia lindleyi (see *Bartonia*)

Mesembryanthemum   Ice Plant, Fig Marigold
 M. cordifolium. "Half hardy. Trailing" (Breck).
 M. crystallinum. "Common Ice Plant—Thick, fleshy leaves . . . have the appearance of being covered with crystals of ice" (Breck).

Mignonette (see *Reseda odorata*)

Mimulus   Monkey Flower
 M. cardinalis. "Very ornamental" (Breck).
 M. intens. Many beautiful varieties produced.
 M. moschatus. Musk plant. "A very pleasing appearance, perfuming the air" (Breck).

Mirabilis   Four-O'clock, Marvel of Peru
 M. jalapa. "Red in its native country, the West Indies. White, yellow, various shades of red, and variegated flowers in the garden. The powder of these roots, washed, scraped and dried, is one of the substances which form the jalap of druggists" (Breck).

Myosotis   Forget-Me-Not
 M. arvensis. "Well known sentimental plant, bearing very delicate blue flowers" (Breck).

Nemophila

   N. atomaria (now *N. menziesii*). Dotted love-grove. "Dotted flower Nemophila—white flowers . . . thickly dotted with purple. All the species are dwarf" (Breck).

   N. insignis. Baby blue-eyes. Showy love-grove. "Opposite-leaved insignias—Brilliant blue flowers" (Breck). Blue love-grove. "Pretty dwarf plant" (Buist).

   N. maculata. "Five-spotted Nemophila—Of great beauty, but not common" (Breck). "Beautiful" (Buist).

Nicotiana longiflora

   "Star Petunia—Pure White" (Breck).

Nigella   Devil-in-a-Bush, Fennel Flower

   N. damascena. "From south of Europe—Light-blue flowers" (Breck). "Love-in-a-Mist" (M'Mahon).

   N. hispanica. "Fennel flower, or devil-in-a-bush (M'Mahon). "Blue and brown" (Buist).

   N. sativa. Nutmeg flower. Black cumin. "Leaves and seed used in cookery. Fine-cut leaves like fennel" (Breck, M'Mahon).

Nolana   Chilean Bellflower

   N. atriplicifolia (now *N. paradoxa*). "Prostrate growth—Deserves a place in every flower-garden. Azure blue with a white centre, the bottom of the flower yellow" (Breck).

Oenothera   Evening Primrose

   O. grandiflora. "Great flowered Evening Primrose—Native of North America. Yellow flowers" (Breck).

   O. longiflora. "Native of Buenos Aires—Uncommonly large and showy yellow flowers" (Breck).

   O. nocturna (now *O. odorata*). "Night-smelling Primrose—From the Cape of Good Hope. Yellow flowers" (Breck).

   O. tetraptera. "White-flowered Evening Primrose—From Mexico. Prostrate-growing . . . with a succession of pure white flowers . . . which make their appearance after the sun has descended the horizon and perish before it rises in the morning" (Breck, Buist).

Papaver

   P. dubium. Smooth poppy (M'Mahon).

   P. rhoeas. Flanders poppy. "Corn Poppy or African Rose—Common weed among grain or gravelly soils in England; but in its double and semi-double varieties it is one of the handsomest of garden annuals . . . sport-

ing in a thousand different varieties. . . . Nothing can exceed the beauty
and delicacy of the flower" (Breck, Bridgeman, M'Mahon).

P. somniferum. "Opium Poppy—Flowers are white, of extra size. Another
variety has dull purple flowers. The double varieties are handsome. . . . A
bed full of full double Poppies, of scarlet, crimson, purple, white and variegated, makes a grand show" (Breck). Common White Poppy (M'Mahon).
"Large, white and scarlet" (Bridgeman).

Pelargonium Geranium

"*Common Scarlet Geranium* a general favorite" (Breck). "*Gold and Silver variegated geraniums*. Flowers pink, carmine, scarlet. Leaves edged with white
and yellow" (Breck). "Horseshoe Geraniums." "Rose geranium and
others. Sweet scented: lemon, musk, etc." (Breck).

Perilla

P. nankinensis. Purple-leaved perilla. "Strange colour of foliage appropriate
for massing in the borders" (Breck).

Petunia (now *P. frutescens* var. *crispa*)

P. nyctaginiflora (now *P. axillaris*). "Large white petunia" (Breck).

P. phoenicia (now *P. violacea*). "Now well known, but not many years as an
inhabitant of our flower-gardens. Flowers purple" (Breck). "From these
two species have been produced innumerable improved varieties in white,
rose, or light-purple, beautifully veined, striped or shaded with crimson
or purple, with dark throats" (Breck). "A beautiful genus of plants of
every variety of color" (Buist).

Phlox drummondii

"First raised from seed from Mr. Drummond in England, named for him
by Dr. Hooker" (Breck). "Worthy of a place in every garden. When
planted in masses, no plant is more showy. Scarlet, crimson, purple,
white and pink, variegated with all intermediate shades. Indispensable"
(Breck).

Platystemon Creamcups

P. californicus. "Of considerable beauty . . . one of the many interesting discoveries of Mr. Douglass, to whom our collections are indebted for its
introduction" (Breck).

Plumbago Leadwort

P. larpentiae (now *Ceratostigma plumbapinoides*). "Bedding plant" (Mead).

Portulaca grandiflora Snowy Portulaca, Moss Rose

"Though one of the most common, still one of the most showy and beautiful annuals." "The double varieties—with the Double Zinnias—are
grand acquisition of the German cultivators" (Hovey's *American Gardener's
Magazine*).

Lychnis chalcedonica.
Pub. by Curtis, *Botanical Magazine,* 1794.

Reseda odorata   Common Mignonette
   "A bed of it should be found in every garden . . . continues to bloom and
   send forth its sweetness all the season, perfuming the whole region about
   the premises. In great demand in and about London . . . creditable London seedsman . . . sold a ton and a half of the seed yearly" (Breck,
   M'Mahon).

Salpiglossis
   S. sinuata. Painted tongue. "Some fine, rich, dark-velvety puce color.
   Extremely beautiful. . . . Some iron-brown, and yellow veined with
   brown. . . . Some pure yellow flowers" (Breck).
Salvia   Sage
   S. coccinea. Texas sage. "Tender smaller scarlet flowers" (Breck).
   S. hispanica. Spanish sage. "Pale blue—Naturalized in West Indies"
   (M'Mahon).

S. splendens. Scarlet sage. "A Mexican plant of extraordinary beauty—large scarlet flowers in spikes" (Breck).

Scabiosa atropurpurea   Sweet Scabious, Mourning Bride
"Suitable for the border. Great variety in flowers . . . some . . . almost black; others a dark puce purple, and various shades, down to lilac" (Breck).

Schizanthus pinnatus   Butterfly Flower
"Beautiful purple and yellow flowers. Common species from which a number of beautiful and improved seedlings have been produced" (Breck).

Schizopetalon walkeri
"Curious white flowers. Native of Chile, whence it was originally introduced by the late Mr. Walker in 1821. Flowers are very fragrant, especially in the evening" (Breck).

Senecio elegans   Ragwort, Double Groundsel
"Double red, crimson white, flesh-colored. . . . Scarcely anything surpassing it for a neat and handsome show" (Breck). Elegant groundsel. Purple Jacobaea. (M'Mahon).

Silene   Campion Catchfly
S. armeria. Sweet william catchfly. Lobel's catchfly. "Dense umbels of pink and another variety with white flowers" (Breck, M'Mahon).
S. compacta. "Compact-flowered—another beautiful species" (Breck).
S. pendula. Nodding catchfly. "Pendulous flowers. Pink flowers" (Breck, M'Mahon).

Silybum   Holy Thistle
S. marianum. "Native of Spain—yellow flowers" (Breck).

Specularia hybrida
"Venus' Looking-glass. Blue flowers which close at the approach of rain and at evenings. . . . Corolla resembles a little round elegant mirror" (Breck).

Sphenogyne speciosa (now *Ursinia anthemoides*)
"Handsome foliage, and a most profuse bloomer. Resemblance to the Calliopsis; rays yellow, disk dark-brown" (Breck).

Tagetes   Marigold
T. erecta. African marigold. "Pale citron-yellow to deep orange" (Breck, Buist).
T. erecta var. *fistulosa*. Quilled African marigold (M'Mahon).
T. erecta var. *flore pleno*. Double African marigold (M'Mahon).
T. patula. French marigold. "One of the old-fashioned flowers. Some of the improved varieties are exceedingly beautiful" (Breck).

Trachymene   Blue Lace Flower, Sky-blue Didiscus (Breck)
T. coerulea. "A handsome annual. Elegant hemispherical heads of the size and shape of a large quilled Aster" (Breck).

Tropaeolum Nasturtium

   T. majus. "Nasturtium—Indian Cress—Rich orange, shaded with crimson
   and various colors. The seeds are used as a substitute for capers, and the
   flowers sometimes eaten as salads" (Breck, M'Mahon).

Verbena

   V. chamaedrifolia (now *V. peruviana*). German-leaved vervain. "Scarlet-
   flowered. Native of Buenos Aires. Introduced into England 1825. The
   dazzling, brilliant, scarlet flowers cannot be exceeded by any plant yet introduced into this country" (Breck).
Viola tricolor (treated by Breck as a biennial or perennial)
   "Heart's Ease, Three Coloured Violet" (M'Mahon).

Xeranthemum annuum Immortelle

   "Purple Everlasting—a variety with white flowers. Valued for their properties of retaining their colors and form, when gathered and dried, and
   much prized in forming winter bouquets for vases, etc." (Breck). "Eternal Flower" (M'Mahon).

Zinnia

   Z. elegans. Youth-and-old-age. "Flowers are handsome when it first com-
   mences the process of blooming" (Breck).
   Z. multiflora (now *Z. peruviana*). Red zinnia (M'Mahon).
   Z. pauciflora (now *Z. peruviana*). Yellow zinnia (M'Mahon).
   Z. tenuiflora (now *Z. peruviana*). Slender-flowered zinnia (M'Mahon).
   "Double Zinnia—great novelty. The cone-forming seed gave it a coarse
   appearance. Later the Double Zinnia transformed this unsightly portion
   into regular petals. . . . Like a well-formed Dahlia" (Breck).

II *Bulbous Roots*

Because many of our authorities were content to treat bulbs, corms, tubers,
and thickened rootstocks in chapters broadly titled "Bulbous Roots," it is not
for us to sort them into today's groupings. Nineteenth-century bulb fanciers
felt it sufficient to divide planting bulbs between those that would do well in
pots and those that would fill the beds with bloom for a season until the annuals could be brought in to take over. Frank Scott, in 1870, announced that

   To make fine display throughout the season for low flowers, it is necessary
   to have at least two sets or crops of plants. One from bulbs such as snow-
   drops, crocuses, jonquils, hyacinths and tulips, all of which may be

planted in October to bloom the following spring, while the bedding plants for later bloom, such as verbenas, portulacas, phlox Drummondii etc. etc. are being started.

At the same time, Fessenden and Henderson, Buist and Bridgeman were sorting out the sorts of "bulbous roots" and "Holland bulbs" that were "well calculated to bloom in a pot."

All approved the common little spring flowers like *Crocus vernus, Scilla siberica,* the two snowdrops (*Galanthus nivalis* and *Leucojum vernum*), and *Muscari,* the grape hyacinth—all of which were hardy and defied "improvement"—and the common hyacinth (*H. orientalis*), whose improvement was to come later. All growers recorded great competition in the "bulbous roots" capable of being forced into named varieties, such as the tulip, the narcissus, and in this case the dahlia. Lilies, like the smaller bulbs, had arrived at a state of acceptance in their historic forms, and the Madonna lily (*Lilium candidum*), Turk's-cap (*L. martagon*), tiger lily (*L. tigrinum*), and the crown imperial (*Fritillaria imperialis*) reigned supreme in many gardens.

The hyacinth was relatively simple before "improvement" into today's fragrant clubs. Henderson lists four kinds of "Roman Hyacinths": Early White (the best); Rose or Redskinned; Blue Roman (not desirable unless as a variety); and Nantes White Spring Hyacinth.

Due to the kindness of the owners of Mountain Shoals in South Carolina, we can list southern standbys, some not hardy farther north: *Jonquilla simplex* and *Narcissus jonquilla; Narcissus polyanthus;* "Double variegated Narcissus"; *Leucojum aestivum* (summer snowflake); *Lycoris radiata* (spider lily), once known as *Nerine japonica; Muscari plumosum,* the "Plumed Hyacinth"; the hardy gladiolus or corn flag, *Gladiolus communis;* and *Bellamcanda chinensis* (blackberry lily).

With the excitement of the discovery that the bulbs brought by traders from Constantinople could be cultivated widely in the fields of Holland to produce all sorts of new combinations of shapes and colors, tulips and narcissus took off as articles of commerce. Even a glance at Parkinson's *Paradisi* of 1629 and Gerard's *Herball* of 1633 will show there was a wealth to choose from. In fact, Parkinson's illustrations can serve as a guide to today's flower shows, as many of his jonquils are still with us under new names. In the nineteenth century fairly broad classifications sufficed.

TULIPS

Bridgeman in 1845 listed the classes of tulips to be exhibited as: Bizarre (yellow ground, marked with purple or scarlet); Flamed Bizarre (a broad stripe up the middle of each petal); Feathered Bizarre (without broad stripe); Byblomen (a white ground marked, as above, with violet or purple); Rose (white ground, marked with rose); and Self (all white or all yellow).

Breck in 1866 added to the same listing Paraquets (fringed crimson and yellow with bright green), "still held in no sort of esteem among florists," and Doubles ("deemed monsters"), and commented that Breeders (raised from seed) are of one color with no variation.

Henderson in 1887, considering tulips chiefly for forcing in pots, listed some, of which two at least are known and popular for planting out today.

For scarlet or red: 'Rembrandt,' 'Artus,' 'Vermilion,' 'Brilliant,' 'Roi Cramoise,' 'Fireflame'

For white: 'Pottebaker,' 'Princess Mary Ann,' 'Queen Victoria,' 'Snowball,' 'White Swan,' 'Grandmaster of Malta'

For yellow: 'Canarybird,' 'Yellow Prince,' 'Duke of Orange,' 'Duchess of Austria,' 'Lucretia'

For rose or pink: 'Cottage Maid,' 'Rosamundi,' 'Rose Adeline,' 'Proserpine,' 'Bride of Haarlem,' 'Everwyn'

For red and yellow striped: 'Duchess de Parma,' 'Kaiserkroon,' 'Queen Emma,' 'Samson,' 'Ma Plus Aimable'

It was estimated by the end of the century that the Linnaean Garden on Long Island was producing 600 varieties of tulips and that an expenditure of $500 would yield a moderate-sized bed of the greater part of the finest varieties.

NARCISSUS

Considering the two opposites from the nineteenth century's rather relaxed attitude toward bulbs in the garden—the extraordinary plenty pictured in Parkinson's early seventeenth-century illustrations and the incredible quantities available today—it is important to realize that Parkinson's climate was much milder than most of America's and that twentieth-century bulbs are much hardier; also, that in the nineteenth century "natural" plantings in woodlands or meadows were still in the future. Jonquils and daffodils came freely into English verse but formally into American planting. Classification by measuring the comparative lengths of the trumpet and the perianth was also to come. In the meantime, everyone knew that, though *Narcissus* covered all, the "daffodil" was *Narcissus pseudo-narcissus* (capable of appearing in many forms, very hardy), the "jonquil" was *narcissus jonquila,* the "Hoop-petticoat Daffodil" was *Narcissus bulbocodium,* the "Poet's Narcissus" was *Narcissus poeticus,* the "Paperwhite Narcissus" was *Narcissus tazeta papyraceus,* and the "Polyanthus Narcissus" was *Narcissus tazeta.* These all figure in nineteenth-century lists. Parkinson's warning that jonquils should never be brought into a room where there are ladies lingers in nineteenth-century warnings about too strong a scent.

Buist, Bridgeman, and Breck listed their preferences in the Narcissus family.

Buist said the narcissus should be treated like a lily. Sorts were cheap and easy to procure. Double and single jonquils could be planted like tulips but must be lifted every third year.

Bridgeman said *Narcissus incomparabilis* is hardy. "Butter-and-eggs" has white petals around the yellow—with a scent some dislike. Grand Monarque de France and Belle Legoise have white flowers with yellow cups. Glorieux has an orange cup on yellow ground. Reine Blanche is white.

Breck, coming later than these two, said of the "Two flowered Narcissus" that Peerless is pale cream with a yellow cup. Of the common daffodil there are many varieties: white flowers with yellow cups; yellow flowers with deep gold cups; double flowers with several cups one within another. The Great Yellow Incomparable is both single and double—the double called "Butter-and-Eggs" by the English, "Orange Phoenix" by the Dutch. The Great Jonquille is yellow with too powerful a scent. The Hoop-petticoat Narcissus (called Medusa's Trumpet in France) has a cup two inches long and broad at the brim, one pale citron, another darker and larger. The white or Poet's Narcissus is snow-white with a pale yellow cup fringed with reddish purple, sweet-scented and sometimes double. The Polyanthus Narcissus is the most desirable but tender, with trusses of six to twenty flowers each.

DAHLIA

The space given to dahlias in nineteenth-century garden books exceeds many times the space given to "Holland bulbs" (tulips, jonquils, hyacinths, etc.). Only fairly recently imported from Mexico, the dahlia was seized upon by all who craved change and success through excess. From a modest, rather tight, round, compact, many-petaled flower on a very long stalk, the dahlia exploded into many markings, shapes, colors, sizes, and forms of petals. By 1853 Thomas Bridgeman listed 160 "Double Dahlias," with descriptions of each—more space than he had given to any other sorts of plants, even annuals and perennials. He noted 100 named varieties carried by Mr. G. G. Thorburn and added that several of the choicest imported seedlings from England cost from fifteen shillings to five pounds each. Our faithful authority, Joseph Breck, cautioned in midcentury that "There is a fashion among amateurs of the floral kingdom, as well as in matters of dress and style, of living among those who lead in fashionable circles of society," that results when a "new flower of fancied merit" is introduced. It can become "all the rage for the time being." As far as the dahlia was concerned, "the time being" lasted for the rest of that century and into the next.

III *Flowers for the House*

Although this small section might have appeared as part of the text, under the heading of "Flower Arranging" or "Indoor Plants," there was very little primary material on this subject to warrant its inclusion. By the end of the century, especially in England, floral paintings and art work for the home were much in evidence, but this did not seem to affect American garden practices. "Window Plantings," too, however popular in Canada, were omitted from the text because they seemed closely dependent upon greenhouses and hothouses and beyond our garden range. However, inquiries have come from expectant modern arrangers about the origins and style of flower arranging and the disposal of blooming, dried, and potted plants in the ordinary house, and it seems only fair to make this brief entry.

We knew dried flowers were popular in moderation on both sides of the Atlantic during the first three centuries of American gardens, and some flowers were especially valued for their drying qualities. These have been included in the text.

As he has had so much to do with the flowers grown in nineteenth-century American gardens, we have kept to Joseph Breck of Boston as our authority. The moderation of his offerings on the subject confirms our feeling that flower arranging as a discipline is a twentieth-century stimulus to American gardening. But Mr. Breck does not stint the house-flower enthusiast. He fairly pours it on. In his *New Book of Flowers,* published in 1866, Breck affords us a quick view of the "Cultivation of Plants in the Parlor." Because, he says, "a few plants in the house are desirable or even indispensable to the female portion of the family, or to invalids who have a taste for flowers," he will list the most advisable sorts for indoor cultivation. With the right exposure and a temperature in the room of 40°F. by night and 60° by day, camellias will do well. With due attention to differing composts and to turning the pots daily, Breck allows the following plants to be grown in the parlor: camellias, roses, pelargoniums, azaleas, ericas, most "new Holland plants," cacti, fuchsias, lemons and oranges ("wash the leaves"), heliotropes, begonias, and salvias. Chrysanthemums are also suitable for "parlour culture," as are the daphne, sweet-scented verbenas, gardenias, azaleas, myrtles, oleanders, pinks, and sweet-scented geraniums. All these, however, will of course do better in a small conservatory from which plants can be introduced to the parlor.

Due to requests from his readers, Breck has a short chapter called "The Art of Constructing Bouquets; Arranging Flowers in Vases, etc." He protests, however, that it is not for him to dictate what is beautiful and pleasing. He takes refuge in quoting from a paper "just in from London" where there is an article called "Flowers and Foreign Flower Fashions" from a gentleman reporting on what he has just seen in Paris. The gentleman has found that "much green with a little color" is the rule. Seldom does one see one color only. Crimson combined with buff, in roses, violet with pink, pale sea green with rose color, are popular, and all or any of these go with white. With flowers in vases, in general, "the great idea in arranging them is to show each flower separately." The reporting traveler abhors "that horrid way, of all others most objectionable, when, having a crowd of flowers, each flower tries to be seen, thus making up a result of a mass of excited petals, like faces turned up in a crowd." The ideal view is to "let each flower repose quietly and calmly upon a bed of green."

"For actual use on the dinner tables," the reporter continues, "the prettiest fashion I have ever seen by far is that of the large open vase supported on gilt branches always so arranged as to look wide and low in proportion to its height . . . filled up with dark green moss . . . raised in the centre, curved upwards. The flowers . . . one of deep red rose, one of the palest blush white, a spray of white convolvulus just touched with pink, a cluster of red drooping flowers (I thought of the rose acacia) one spray of pale wild rose, one bright pink rose, a cluster of white acacia, and a drooping branch of the pink convolvulus . . . colors all shades of rose and white . . . each flower simply laid on the green, the flowers just touching, each with its own green leaves. No attempt to fill up the center, stems hidden in the moss."

On a larger scale, going in to dinner this gentleman walked under a vase standing on a pedestal where, again, each flower was allowed to show itself distinctly. This time the arrangement was of "gigantic" flowers: "sunflowers and dahlias, great roses and gladioli, each with large green leaves and long stems—beautiful in its strangeness."

And there is our Mr. Breck, tactfully avoiding local involvement.

Incidentally—as we seem to say when something of importance to our cause seems to turn up on its own initiative—if Mr. Breck has only the above to say and if none of our other worthies devotes any space to growing cut flowers for indoor decoration, we can judge the relative importance of nineteenth-century flower arranging before it achieved the eminence it commands today.

Mrs. Loudon spoke kindly, before the middle of the century, of a few flowers in fresh water for a desk or a table. It is almost as if to match her assessment of flowers for the English house as an art form or a direct influence upon gardening that we append here a contemporary view of mid-nineteenth-century Boston's involvement in the future competitive sport.

A famous French silhouette artist, A. Edouart, made a popular and ob-

Silhouette of the Paige family, by A. Edouart.

viously profitable visit to Boston recording studies of the first families in their drawing rooms. Here we have his portrait of the Paige family in 1842 in their drawing room. Each of the four members is engaged in a commendable occupation. The little girl is studying her music. The little boy (who grew up to be a collector of rare art) considers his bow and arrows. Mr. Paige sits at his desk, and Mrs. Paige is arranging a few cut flowers in a vase.

This is an appropriate grace note to signal the extent of the nineteenth century's involvement in the charms of arranging flowers before it became a modern industry, fraught with rules and powered by competition.

William Cobbett, in his *American Gardener*—at least in our edition of 1821— gives credence to our feeling that flowers for the house in the early nineteenth century were a continuation of the eighteenth century's devotion to large pots of fragrant shrubby plants like rosemary and lavender. Where a greenhouse is lacking he recommends showy and fragrant shrubby plants in large pots to keep the house attractive in winter.

IV *Herbs*

The cultivation of special plants for flavoring, coloring, scenting, and dosing is as old as time. Their segregation to special beds in the garden, in specially

designed artistic layouts, is comparatively modern. Parkinson's 1629 suggestions had been to incorporate useful herbs into the vegetable garden, with perhaps the exception of those for medicinal purposes, which might be arranged
in their own section to make them easy to find in emergencies. Using plants of
attractively varied leaf colors and compact habits to keep the edges of raised
beds "in fashion" was time-honored but derived from their appearance and response to constant shearing rather than from their uses. However, M'Mahon
made great play of them in his monumental *American Gardener's Calender* of
1806 by including practically all herbs ever mentioned elsewhere. It may have
been this section that engendered doubts more than a hundred years later as to
whether this great man's great nursery would really have supplied all the plants
mentioned in the great book. The more we garden, the more we may wonder.
However, when practical and less inspired seedsmen and nurserymen in the
nineteenth century list a modest number of these useful plants we include them
here as being indeed part of the American nineteenth-century garden.

We give M'Mahon's "List of Aromatic, Pot, and Sweet Herbs" exactly as it is
printed in his Catalogue in *The American Gardener's Calendar,* 1806.

Anise   *Pimpernella Anisum*
Basil, sweet   *Ocymum basilium*
Basil, bush   *Ocymum minimum*
Caraway   *Carum Carvi*
Clary   *Salvia sclarea*
Coriander   *Coriandrum sativum*
Chamomile   *Anthemis nobilis*
Dill   *Peucedanum graveolens*
Fennel, common   *Foeniculum vulgare*
Fennel, sweet   *Foeniculum dulce*   Florence fennel
Hyssop   *Hyssopus officinalis*
Lavender   *Lavendula vera* English   *Lavendula spica* French
Lovage   *Levisticum officinale*
Marigold, pot   *Calendula officinalis*
Marjoram, sweet   *Origanum marjorana*
Marjoram, pot   *Origanum Onites*
Marjoram, winter sweet   *Origanum Heracleoticum*
Mint, spear   *Mentha viridis*
Mint, pepper   *Mentha piperita*
Mint, pennyroyal   *Mentha Pulegium*
Mint, horse   *Monarda punctata*
Rosemary   *Rosmarinus officinalis*
Sage, common   *Salvia officinalis*

Savory, summer   *Satureia hortensis*
Savory, winter   *Satureia montana*
Smallage   *Apium graveolens*
Tarragon   *Artemisia Dracunculus*
Thyme, common   *Thymus vulgaris*
Thyme, lemon   *Thymus serpyllum*

M'Mahon's "Plants Cultivated for Medicinal Purposes"

Ague-weed, Thoroughwort   *Eupatorium perfoliatum*
Angelica, garden   *Angelica Archangelica*
Betony, Wood   *Betonica officinalis*
Bugloss   *Anchusa officinalis*
Carduus benedictus   *Centaurea benedicta   Carbenia benedicta*
Celandine   *Chelidonum majus*
Comfrey, common   *Symphytum officinale*
Cucumber, bitter   *Cucumis Colocinthus   Citrullus colocynthis*
Elecampane   *Inula Helenium*
Flax, common   *Linum usitatissimum*
Fenugreek   *Trigonella Foenum Graecum*
Feverfew   *Matricaria parthenium   Chrysanthemum Parthenium*
Foxglove   *Digitalis purpurea*
Gromwell   *Lithospermum officinale*
Hemlock   *Conium maculatum*
Horehound   *Marrubium vulgare*
Hound's-Tongue   *Cynoglossum officinale*
Liquorice   *Glycyrrhiza glabra*
Madder, Dyer's   *Rubia tinctorum*
Mallow, marsh   *Althaea officinalis*
Mugwort, Common   *Artemisia vulgaris*
Nep, or Catmint   *Nepeta Cataria*
Nettle, Stinging   *Urtica urens*
Palma Christi, or Castor Oil Nut   *Ricinus communis*
Pimpernell   *Anagallis arvensis*
Pink-root, Carolina   *Spigelia marilandica*
Poppy, opium   *Papaver somniferum*
Rue, garden   *Ruta graveolens*
Rhubarb, true turkey   *Rheum palmatum*
Rhubarb, common   *Rheum Rhaponticum*
Scurvey-grass   *Cochlearia officinalis*
Snake-root, Virginia   *Aristolochia serpentaria*
Southernwood   *Artemisia Arbrotanum*

Tansey   *Tanacetum vulgare*
Tobacco, cultivated   *Nicotiana Tabacum*
Tobacco, common English   *Nicotiana rusticum*
Weld, Woad, Dyer's Weed   *Reseda Luteola*
Winter Cherry   *Physalis Alkekengi*
Wormseed, goosefoot   *Chenopodium anthelminticum*
Wormwood   *Artemisia Absinthium*
Yarrow   *Achillea Millefolium*
Yarrow, sweet, or milfoil   *Achillea Ageratum*

These are M'Mahon's names as he gave them. The corrections are from Grieve's *Modern Herbal*.

Bridgeman, in 1847, in his *Young Gardener's Assistant,* presents two lists like M'Mahon's of "Aromatic, Pot, and Sweet Herbs" and "Plants Cultivated for Medicinal and Other Purposes" except that Bridgeman, in a gesture of elegance, adds to each its title in French—"Graines d'Herbes Aromatiques, Odiferantes et à l'Usage de la Cuisine"—after which he abandons the second language and makes his lists of the common English name followed by the botanical.

Angelica he allows to be culinary, followed by Anise and Basil. Then: Borage (*Borago officinalis*) and Burnet (*Poterium sanguisorba*); Caraway; Chervil (*Scandix odorata cerefolium*); Clary; Coriander; Dill; Fennel; Marigold; Marjoram, sweet; three mints—Spear, Pepper, and Pennyroyal; Sage; Savory, summer, and winter; Tarragon; Thyme, common and lemon.

Of those mentioned by M'Mahon, the only ones Bridgeman left out are Chamomile, Hyssop, Lavender, and Rosemary.

By the time Henderson wrote his masterful *Handbook of Plants* in 1881, all the above "herbs" are listed in their alphabetical order as plants but are not singled out or grouped, and the definition of *herb* is merely "A plant that does not possess a woody stem."

So we must wait for the next century for a "herb garden."

V   *Perennials*

Of perennial plants, the gardener's staunchest friends and standbys, what can one say in nineteenth-century terms without reflecting today's gratitude and dependence? In their various roles through the centuries, they have stood by us to a degree where the owner-maintained garden today would be bereft without them. With many hands to do the work in the nineteenth century, they may have ranked as only equal with annuals except in the borders. Today they are indispensable and need only to be called upon in their vast variety.

Achillea Yarrow, Sneezewort
  A. aurea (now *A. tomentosa*). "Rich yellow, not so hardy" (Breck).
  A. millefolium. "Yarrow, native, pink variety, desirable" (Breck).
  A. ptarmica. "Sneezewort . . . desirable border flower, especially double.
  White" (Breck). "Double milfoil, white" (Downing).
Aconitum Monkshood, Wolfsbane
  A. japonicum. "dark blue. July, August."
  A. napellus. "common one, poison. Monkshood, purple."
  A. rostratum. "beautiful purple."
  A. variegatum. "beautiful, light blue edged white. July, August." "Purple
  and white. Variegated" (Breck, Bridgeman, Downing).
Actaea Baneberry, Cohosh
  A. alba. "white berries."
  A. rubra. "red berries—Indigenous perennial, suitable for shrubbery. Found
  in woods. Both varieties have white flowers" (Breck).
Adonis vernalis Pheasant's-Eye
  "Spring-flowering, yellow." "Fine border flower . . . any common soil,
  flowers large yellow-rayed . . . about twelve petals, leaves much divided"
  (Buist). "Handsome perennial border plant. Native south of Europe. Any
  common soil if not too heavy" (Breck, Bridgeman, Downing).
Agrostemma Rose Campion, Mullein Pink
  "Common showy border flower, easily kept by dividing the roots. Deep
  red, white, white with pink centre" (Breck).
  A. flos-cuculi (now *Lychnis flos-cuculi*). Ragged robin, or red lychnis.
  A. coronaria, rosea, alba, etc. (now *L. coronaria*). Rose campion or Mullein
  pink. Rose, white, etc. (Bridgeman).
Alcea (see *Althaea*)
Althaea (now *Alcea*) Hollyhock
  A. rosea. "Chinese Hollyhock. Cultivated since 1830. Many prize winners
  now" (Breck). "Antwerp, China and English Hollyhocks—various col-
  ors. *A. rosea, A. chinensis, A. anglica*" (Bridgeman).
Althaea officinalis Mallow
  "Althea" common name for *Hibiscus syriacus*.
Alyssum saxatile (now *Aurinia saxatile*)
  "Rock or Golden Alyssum, desirable vernal flower. Plant with *Phlox stoloni-
  fera* with red flowers and *Phlox subulata* pink or white" (Breck). "Golden
  basket, yellow. April" (Bridgeman, Downing).
Amsonia Blue Star
  A. tabernaemontana. "Hardy natives of southern states, pretty blue flowers.
  June" (Breck).

Anchusa   Bugloss, Alkanet

A. *italica* (now A. *azurea*). "Small brilliant flowers."

A. *officinalis*. "Common bugloss. Blue" (Breck, Downing).

Anemone   Windflower

A. *coronaria*. "Windflower, various colors. Mediterranean" (Bridgeman).

A *hortensis*. "Garden anemone. Species from which florists' flower originated" (Breck).

A. *nemorosa*. "Wood anemone, European" (Breck).

A. *patens* var. *nuttalliana*. Gray. American pasque flowers.

A. *pulsatilla*. "Pasque flower of Europe. Described by Gerard" (Breck). "Pasque flower, blue" (Downing).

A. *thalictroides* (now *Anemonella thalictroides*). "Double wood anemone. Rue anemone" (Downing).

Anemonella thalictroides (see under *Anemone thalictroides*)

Anthericum   Spider Plant

A. *liliago*. "St. Bernard's Lily. Yellow. Any common soil. June, July, August" (Buist).

Antirrhinum   Snapdragon

A. *majus*. "Common or large Snapdragon" (Breck). "Red and white Snapdragon" (Downing). "Curious, ornamental, mostly perennials or biennials. Sported into many varieties" (Buist, Bridgeman).

A. *vulgaris* (now *Linaria vulgaris*). Toadflax. (Our native snapdragon was formerly cultivated.) (Breck, Buist, Bridgeman).

Aquilegia   Columbine

A. *alpina*. "Alpine Columbine, purple" (Bridgeman).

A. *canadensis*. Wild Columbine, Scarlet. May" (Downing).

A. *glandulosa*. "Newly introduced, dwarfish, sky-blue, white margin to corolla" (Breck).

A. *vulgaris*. "Columbine, various colors" (Bridgeman).

Aruncus sylvester (now A. *dioicus*)   Goatsbeard (Breck)

Asclepias   Butterfly Weed, Milkweed

A. *syriaca*. "Milkweed, silkweed" (Breck).

A. *tuberosa*. Pleurisy root. "Finest genus of all. Orange, dry situations" (Buist). "Butterfly-weed" (Breck).

Asphodelus

A. *luteus* (now *Asphodeline lutea*). King's spear (see Jackson's Hermitage, chap. 14).

A. *ramosus*. "Branchy asphodel, white" (Downing).

Aster   Starflower, Michaelmas Daisy

A. *cordifolius*. Blue wood aster.

A. corymbosus (now *A. divaricatus*). White wood aster (Breck).
A. linearifolius. "Fine-leaved aster. White" (Downing).
A. macrophyllus. "Broad-leaved aster. White" (Downing).
A. multiflorus (now *A. ericoides*). "Snow white, 2'" (Breck).
A. novae-angliae. "3'–4', large purple flowers" (Breck).
A. novi-belgii.
A. puniceus. "Brilliant light blue. 3'–4'" (Breck).
Aureolaria flava; Aureolaria virginica (for comments, see *Gerardia*)
Aurinia saxatilis Basket-of-Gold (for comments, see *Alyssum saxatile*)

Baptisia australis
"False Indigo. A blue dye from the leaves." "Handsome border plant" (Breck, Henderson).
Bellis perennis hortensis English Daisy
"Common daisy. March to August, 3' tender" (Breck). "Garden daisy, various colors, tender" (Bridgeman). "Many double varieties. 'Crown,' 'Carnation' red and white petals, very double, twice size of other" (Buist). "Daisy. Small delicate perennial well calculated for pots placed in the sitting room" (Henderson).

Caltha
C. palustris flore pleno. "A good border plant, damp place, April–June" (Buist). "Double marsh marigold. Yellow" (Downing).
Campanula Bellflower
C. carpatica. Tussock bellflower. "Carpathian harebell" (Henderson, Breck).
C. glomerata. Clustered bellflower. "Siberia. 2'" (Breck).
C. grandiflora (now *C. medium*). "Great-flowered Harebell and others" (Henderson, Buist, Breck, Downing). (see also *Platycodon grandiflorus.*)
C. lactiflora (Breck).
C. macrantha. "Large blue. July. Variety of *C. persicifolia*" (Downing).
C. medium. "Canterbury Bells. August, September" (Breck). "Blue, white, biennial" (Bridgeman).
C. persicifolia. "Peach-leaved Harebell" (Henderson). "Blue, white" (Bridgeman). "Peach-leaved. One of the finest" (Breck, Buist).
C. persicifolia alba plena.
C. pyramidalis. "Pyramidal Bellflower. Chimney Bellflower. Blue, tender, biennial." "4'–6'." "August" (Bridgeman, Breck, Downing).
C. rapunculus. Biennial. "Rampion" (Breck).
C. rapunculoides. "Nodding, blue. June. Roving Bellflowers" (Downing).
C. speciosa (now *C. glomerata* var. *Dahurica*). (Breck, Buist).

Campanula rotundifolia,
harebell, from *Breck's New
Book of Flowers,* 1866.

C. trachelium. "Double blue and white. July" (Downing). "Throatwort" (Breck).

C. urticifolia (now *C. trachelium*). "White, also double variety" (Buist).

C. versicolor. "Many ornamental plants for the flower garden and shrubbery. Agree better with our climate than that of Europe" (Buist). "Bell flower, our native, grows beside the Merrimack, and is like *C. rotundifolia* from England" (Breck). (*C. rotundifolia* is the common harebell.)

Cassia marilandica
   "Wild Senns—Hardy, Indigenous, Yellow" (Breck, Henderson, Bridgeman, Downing).

Catananche caerulea   Cupid's Dart
   "Blue Catananche. Handsome. From Europe. Brilliant blue, 1 1/2'. July and August" (Breck, Downing).

   Cedronella triphylla (now *C. canariensis*)   Balm of Gilead, Canary Balm (Bridgeman).

Centranthus ruber (for comments, see *Valeriana*)

Cheiranthus Wallflower

   C. cheiri. "Tender. Were it perfectly hardy, would be more highly esteemed. Yellow and orange" (Bridgeman). "Survives the mildest of our winters" (Buist).

Chelone Turtlehead, Snakehead

   C. barbata (now *Penstemon barbatus*—Nuttall). "Bearded chelone. Orange" (Downing).

   C. glabra. "Turtlehead. White, rose or purple. Entirely North American. Astonishing not more sought after" (Breck, Buist).

   C. obliqua (Buist).

Chrysanthemum

   C. frutescens. Paris daisy.

   C. indicum. "Chinese Chrysanthemum. Handsomest autumnal flower, easily cultivated. Varieties endless: early, late, tassel-flowered, quilled, flat-petalled, etc. Every shade: light purple, yellow, white, lilac, blush brown, redbrown, etc." (Breck). "The grand hybrid varieties of the Chrysanthemum now run into thousands of almost every shade except blue and bright scarlet" (Henderson).

   C. leucanthemum. Oxeye daisy.

   C. macrophyllum. Tansy.

   C. parthenium. "Feverfew. Resembles Chamomile" (Breck).

Cimicifuga racemosa Bugbane

   "Black Snake-root. Black Cohosh" (Nuttall). "Native. Worth growing where there is room" (Breck).

Convallaria Lily of the Valley, Solomon's Seal

   C. majalis. "Lily of the Valley. White" (Breck, Bridgeman, Downing, Henderson).

Coreopsis

   C. drummondii (now *G. basalis*). Golden wave.

   C. grandiflora. "September" (Downing).

   C. lanceolata. "Yellow, compound, early, flowers all summer. 2′" (Breck).

   C. tenuifolia. "Like *C. verticillata* but yellow disk. Suitable for front of border. 1′ July and August" (Breck, Buist, Downing).

   C. tripteria. "Finest of the genus and will grow in any soil" (Buist). "Suitable for the shrubbery. 6′" (Breck).

   C. verticillata. "Pale yellow, brown centre, flowers star-shaped. July, October" (Buist).

   (Bridgeman places *Coreopsis* under "*Calliopsis grandiflora, lanceolatam auriculata,* etc.")

The Chrysanthemum: another garden standby from
the Orient. Pub. by Curtis, *Botanical Magazine*,
1796.

Coronilla   Crown Vetch
  C. varia. "Pink garlands. Worthy if annually dug around to control spread-
    ing" (Breck).
Corydalis   Fumitory
  C. aurea. Yellow.
  C. bulbosa. "Purple Fumewort" (Breck).
Cymbalaria muralis. "Kenilworth Ivy" (Bridgeman).

Delphinium
- *D. breckii.* "Double blue" (Breck).
- *D. chinense* (now *D. grandiflorum*) (Downing).
- *D. elatum.* "Bee Larkspur. Covered with soft down. 5'–7'" (Breck, Buist). "Bee Larkspur. Blue and Brown" (Bridgeman).
- *D. grandiflorum* (var. *chinense*). "The best" (Henderson, Buist). "Blue, white, pink" (Bridgeman). "2'–3', June–October (Breck). "Double Chinese Perennial Larkspur. Most magnificent of herbaceous plants. Propagation only by root division. Beautiful blue. June–October, 3'" (Breck).
- *D. nudicaule.* "Scarlet Larkspur" (Henderson).
- *D. speciosum.* "Showy Larkspur" (Downing).

Dianthus   Pink
- *D. alpinus.* "Dwarf Pink" (Breck, Buist).
- *D. barbatus.* "Sweet William, various colors" (Bridgeman, Breck, Buist).
- *D. caryophyllus.* Clove pink. "Carnation Pink. No flower more desirable. Various form." Flakes (two colors only, stripes) (Breck). Bizarres (variegated, spots, stripes, at least three colors). Picotees (white ground spotted with red, purple or other colors) (Breck). "Carnation Pink, various colors, tender" (Bridgeman, Buist, Downing).
- *D. chinensis.* "China Pink. No fragrance" (Breck). "Indian Pinks. Variegated" (Downing). "Chinese Imperial Pinks. Variegated" (Bridgeman, Buist). (Annual, biennial, and short-lived perennial.)
- *D. deltoides.* Maiden pink. "Mountain Pink. Red. July" (Downing). "London Pride, variegated" (Bridgeman).
- *D. fragrans* (similar to *D. plumarius*). (Buist).
- *D. hortensis.* "Garden Pink. Last of June" (Breck). "Garden Pink, many sorts and colors" (Downing). "Clove Imperial Pink, crimson" (Bridgeman).
- *D. plumarius.* Grass pink. "Pheasant-eyed pink, variegated" (Buist). "Feathered Pink" (Breck).
- *D. superbus.* "Fringed Pink" (Henderson). "Superb Pink" (Breck, Buist).

Dicentra   Bleeding Heart
- *D. canadensis.* Squirrel corn (Breck).
- *D. cucullaria.* Dutchman's breeches (Breck).
- *D. eximia.* "Red Flowered Dicentra. Plumy dicentra" (Breck).
- *D. spectabilis.* Bleeding heart (Henderson).

Dictamnus   Gas Plant, Fraxinella
- *D. alba.* "White, June. Called Fraxinella because of ashlike leaves. From Germany. May–June 3'" (Breck). "Sandy loam. May, June" (Buist). "Purple fraxinella" (Downing, Henderson).
- *D. rubra, alba.* "Fraxinella, red, white. 1'–3'" (Bridgeman).

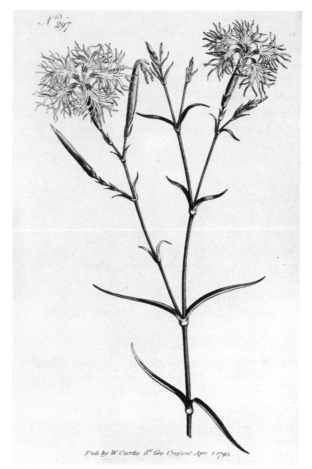

Dianthus superbus. Pub. by Curtis, *Botanical Magazine,* 1795.

Digitalis Foxglove
  D. alba. "Biennial, White" (Breck, Buist, Bridgeman).
  D. lutea. "Small yellow foxglove. 2'. July–August" (Breck).
  D. purpurea. Common Foxglove. "Foxglove, purple. Biennial" (Breck,
    Bridgeman, Buist). "Handsome flowers, purple the most beautiful . . .
    possesses high medicinal virtues" (Fessenden).
Dodecatheon meadia Shooting Star
  "American cowslip. Lilac" (Downing). "Loam, half shady, damp" (Buist).

Dracocephalum Dragon's Head.
  D. *canariense* (now *Cedronella triphylla*). "Balm of Gilead. Fragrant tender" (Bridgeman).
  D. *dentatum* (now *Physostegia denticulata*). "Pink, smaller than D. virginianum. 2'" (Breck).
  D. *grandiflorum*. "Purple, July" (Downing).
  D. *sibiricum* (now *Nepeta sibirica*). "Light blue. 1'" (Breck).
  D. *virginianum* (now *Physostegia virginiana*). "Bluish pink, 3'-4'" (Breck, Buist).

Echinacea purpurea Purple Rudbeckia (Bridgeman).
Echinops Globe Thistle
  E. *ritro*. "Blue, Small" (Downing).
  E. *sphaerocephalus*. "Purple, Great" (Bridgeman).
Epilobium Great Willow Herb, Fireweed
  E. *angustifolium*. "Good under tree drips" (Breck). "Purple, August" (Downing).
Erica carnea Winter Heath (Henderson)
Eschscholzia californica California poppy (Breck). Golden Eschscholzia (Downing)
Eupatorium Thoroughwort, Boneset, Joe-Pye weed
  E. *aromaticum*. "White, fragrant" (Downing). "August–October" (Buist).
  E. *coelestinum*. Mist flower. "Sky-blue Eupatorium" (Breck). "Desirable. September–November" (Buist). "Azure blue" (Breck).
  E. *perfoliatum*. "Boneset" (Breck).
  E. *purpureum*. Joe-Pye weed. "Purple Thoroughwort. August 4'-6'" (Breck).
Euphorbia Spurge
  E. *corollata*. Flowering spurge. "Great flowered" (Breck).

Filipendula Meadowsweet, Dropwort
  F. *hexapetala* (now *F. vulgaris*). Dropwort.
  F. *palmata* (Breck).
  F. *ulmaria pleno*. Queen of the meadow (Breck).
Foxglove (see *Digitalis*)
Funkia (now all *Hosta*) Plantain Lily, Day Lily
  F. *caerulea* (now *Hosta ventricosa*). "Beautiful with blue flowers" (Buist).
  F. *japonica* (now *H. plantaginea*). "Fragrant" (Buist). "Japan Plantain Lily" (Henderson).
  F. *ovata* (now *H. ventricosa*). "Blue Plantain Lily" (Henderson).
  F. *sieboldii* (now *H. sieboldiana*). "Lilac Funkia" (Downing).

F. variegata (now *H. undulata*). "Striped flowers. Very rare, but recently introduced from Japan" (Buist).

Gaillardia   Blanket Flower
  G. aristata. "Bristly Gaillardia. Yellow. June" (Downing).
  G. bicolor. "Orange Gaillardia. Tender" (Downing).
  G. picta. "Handsome plant, beautiful crimson flowers 2″ across. Each petal tipped with yellow" (Breck).
Gentiana
  G. acaulis. "Pretty dwarf for edgings. Not perfectly tried yet" (Buist).
  G. ascendens. "Porcelain-flowered" (Bridgeman).
  G. catesbaei. Sampson's snakeroot (Buist).
  G. crinita (now *Gentianopsis crinita*). "Fringed Gentian. Propagate from seed. September. Exceeded by few native plants" (Breck, Buist).
  G. lutea. "Yellow" (Bridgeman).
  G. ochroleuca (now *G. villosa*). Sampson's snakeroot (Buist).
  G. purpurea. "Purple" (Bridgeman, Buist).
  G. saponaria. Soapwort. "Barrel-flowered Gentian. By streams, may be transplanted without difficulty" (Breck).
  G. septemfida. Crested gentian. "Very showy, abundant, flowering generally blue, few yellow, some white. Light, rich soil" (Buist).
Geranium. Cranesbill
  G. antulatum. "Angular-stalked Cranesbill, native of Europe, cultivated since 1789. Pink flowers in June" (Breck).
  G. maculatum. "Cranesbill. Native" (Breck).
  G. pratense. "Crow-foot-leaved" (Breck).
  G. sanguineum. "Bloody Geranium. Red" (Downing).
  G. sanguineum var. *lancastrense.* "Purple, dwarf, creeping" (Breck).
Gerardia
  G. flava (now *Aureolaria flava*). "American Foxglove" (Breck).
  G. quercifolia (now *Aureolaria virginica*). "Difficult to transplant, possibly may be grown from seed" (Breck, Downing, Buist).
Goniolimon speciosum; G. Tatanicum (for comments, see *Statice*)
Gypsophila paniculata   Baby's Breath
  "Panicled gypsophila" (Henderson).

Helianthus   Sunflower
  H. altissimus. "Sunflower, yellow, 3′–4′″" (Bridgeman).
  H. multiflorus (now *H. decapetalus*). "Double. Fine plant to beautify a border" (Breck, Bridgeman). "The double flowered variety . . . the size and form of a Dahlia" (Breck).

(Breck says a number of indigenous perennial sunflowers are not to be tolerated in the garden.)

Hemerocallis Day Lily

  H. flava (now *H. lilioasphodelus*). Lemon lily. "Early, June" (Breck).

  H. fulva. Orange day lily. "Copper" (Breck, Buist).

  H. graminea (now *H. minor*). "Yellow, copper" (Buist).

Hepatica Liverwort, Liverleaf

  H. acutiloba. "Less common" (Breck).

  H. triloba (now *H. americana*). "Great favorite in the flower border. There are some lively cultivated double blue, red and white, but rarely seen" (Breck). "Hepatica, blue, pink, tender" (Bridgeman). "Double hepatica, blue" (Downing).

Hesperis Sweet Rocket

  H. matronalis. "Garden Rocket. Single varieties common in most gardens. Double variety superb" (Breck). "Garden Rocket, purple" (Bridgeman, Downing).

Hibiscus Rose Mallow

  H. grandiflorus. Great rose mallow (Buist).

Hibiscus moscheutos, Breck's New Book of Flowers, 1866.

H. militaris. "Halbert-leaved Rose-Mallow" (Breck).

H. moscheutos. "Swamp Rose-Mallow" (Breck).

H. palustris (now *H. moscheutos*) (Bridgeman, Buist).

H. roseus (now *H. moscheutos*) (Buist).

H. speciosus. "Large white with deep red centers" (Buist, Bridgeman).

Hollyhock (see *Alcea*)

Hosta Plantain Lily, Japan Day Lily (for comments, see *Funkia*)

H. plantaginea (Breck, Downing).

Houstonia coerulea. "Blue Houstonia for May bloom" (Downing).

Hypericum Saint-John's-Wort

H. ascyroides. "Giant Hypericum. Likes shade" (Breck).

H. calycinum. Rose of Sharon, Aaron's beard. "Large yellow flowers" (Breck).

Iberis Candytuft, Evergreen Iberis, Edging

I. sempervirens. "Deserving a place in the garden" (Breck).

I. tenoreana (now *I. pruitii*). Mat-forming. Similar to *I. sempervirens*" (Breck).

Iris

I. cristata. "Crested Iris. Light blue, very desirable, 3"–6"." Dwarf (Breck, Buist).

I. florentina (now *I. germanica* var. *florentina*) "Florentine Iris. Large, white, 2', May June" (Buist). "Orris root of druggists" (Buist, Downing).

I. furcata (now *I. aphylla*) (Buist).

I. germanica. "German Iris. Common fleur-de-lis of garden. Blue and purple. 2', May and June" (Breck, Buist).

I. gracilis (now *I. prismatica* and *I. pumila*). "Another native, not common" (Breck).

I. nepalensis (now *I. germanica* var. *nepalensis*) (Buist).

I. pallida. Orris "fragrant" (Buist).

I. pseudacorus. "Yellow English water Iris" (Breck).

I. pumila. "Dwarf Iris. Purple, 3" 6", May" (Breck).

I. sambucina. "Elder-scented Iris. Beautiful pale blue, many flowered. Variegated, grasslike foliage, roots fibrous, flowers rise above foliage, 4', late June" (Breck).

I. sibirica. "Siberian Iris, blue, May" (Downing).

I. susiana. Mourning iris. "Also *I. chalcedonia*. Most beautiful tuberous, not bulbous" (Breck). "Chalcedonian Iris, mottled. June" (Downing). "Finest, may be tender, brown spotted" (Buist).

I. variegata. "Delicate, elegant, bluish white" (Breck).

I. verna. Dwarf (Buist).

I. versicolor. "Blue flag" (Breck).

Iris. Pub. by Curtis, *Botanical Magazine*.

I. virginica. "Slender Blue Flag" (Breck).
I. xiphium. Spanish iris (Breck).

Jeffersonia diphylla   Twinleaf
   "Two-leaved Jeffersonia, white" (Downing).

Lamium rugosum (now *L. album*)   Dead Nettle
   "Rough-leaved Lamium. Curious white flowers, suitable for rock-work.
      Unpleasant odor. Names for sea monster" (Breck).
Lathyrus   Everlasting Pea
   L. grandiflorus. Two-flowered pea. "Large rose-coloured" (Breck).
   L. latifolius. "Purple pink in clusters" (Breck). Perennial pea (Downing).
   L. luteus. "Handsomest of papilionaceous tribe" (Breck).

L. niger. "Dark purple vetch. Hardy. Pretty. June, July" (Downing).

Liatris Blazing Star, Gayfeather

L. elegans. "Elegant Blazing Star, purple" (Downing, Buist, Bridgeman).

L. macrostachya (now *L. spicata*). "Strong heavy soil. Blooms first from the extremity, unlike other spike-flowering genera" (Buist).

L. scariosa. "Blue Blazing Star" (Downing).

L. spicata. "Long, spiked purple liatris" (Bridgeman).

L. squarrosa. "Large purple heads" (Buist). "Blazing Star, blue" (Downing).

Linum perenne Flax

"Bright blue flowers. Long time in flower" (Breck). "Perennial Flax, purple" (Bridgeman). "Perennial Flax, blue, June" (Downing).

Lobelia

L. cardinalis. "Cardinal Flower, red, July" (Downing). "Cardinal flowers, in

Lobelia cardinalis, Breck's New Book of Flowers, 1866.

LOBELIA CARDINALIS.

varieties, scarlet" (Bridgeman). "Introduced to England in 1649" (Breck).

L. fulgens (now *L. splendens*). "From Mexico in 1809, scarlet" (Breck).

L. siphilitica. Great blue cardinal flower. "Native of Virginia. Introduced to England in 1665. Indian remedy. Sir William Johnson purchased secret from Indians" (Breck).

L. spicata. "Pale blue, roadsides" (Breck).

L. splendens. "From Mexico in 1814. Scarlet" (Breck).

Lunaria Honesty (see also under "Annuals")

L. biennis. "Honesty. Satin Flower, lush biennial" (Bridgeman, Breck).

Lupinus

L. grandiflora (now *L. polyphyllis*) (Breck).

L. nootkatensis. "Nootka Sound Lupine, blue" (Downing).

L. perennis. "Indigenous to New England. Blooms with *Silene pensylvanica* and *Viola pedata*. Through the exertions of Mr. Douglas on the Northwest coast, many added" (Breck). "Perennial, blue" (Bridgeman).

L. polyphyllus. "Purple Lupine" (Downing, Breck).

Lychnis Maltese Cross, Mullein Pink

L. alpina. "Variegated, Mountain Lychnis" (Bridgeman).

L. chalcedonica. Maltese cross. "Double Scarlet Lychnis. Splendid. Common in most gardens under London Pride" (Breck). "Frequently lift and replant or it dwindles" (Buist). "July" (Downing). "Scarlet" (Bridgeman).

L. coronaria. "Rose Campion, Mullein Pink" (Breck).

L. coronata. "Showy, from China" (Breck). "Chinese orange Lychnis, August" (Downing).

L. dioica. "Purple" (Breck).

L. flos-cuculi. "Ragged Robin" (Breck). "Ragged Robin or French Pink" (Buist).

L. flos cuculi pl. "Double Ragged Robin, red" (Downing).

L. flos-Jovis. Flower of Jove. "Red flowers, downy leaves" (Breck).

L. fulgens. "Splendid from Siberia. 1 1/2'." (Breck). "Frequently lift and replant or it dwindles" (Buist). "Red, June" (Downing).

L. haageana. "Large scarlet flowers" (Breck).

L. viscaria. "Pink. German Catchfly" (Breck).

L. viscaria plena. "Fine border flower" (Breck, Downing).

Lysimachia nummularia

"Moneywort. Most are bog plants, easy to cultivate" (Breck). "Trailing Loosestrife, yellow" (Downing).

Lythrum Loosestrife

L. alatum (Buist).

L. salicaria. "British native. Handsome border flower. Purple spikes. 3'–4',

July, August" (Breck). "Willow Herb, purple" (Downing).
L. virgatum (Buist).

Malva   Mallow, Musk Mallow
  M. alcea. "Purple. From Germany, July–October. 3', Hardy. Varieties of
    pink and white . . . Lower leaves angular, upper five-parted" (Breck).
  M. moschata. Muskmallow.
Matthiola   Stock
  M. incana. Brompton stock. "Queen-stock and its varieties require frames in
    winter" (Buist). "Common stock, Queen stock" (Breck, Bridgeman).
    "Brompton-stock and its varieties require frames in winter" (Buist).
    "Stock-gilly, indispensable but tender" (Buist). "Stock Gillyflower, nu-
    merous varieties, scarlet, white, purple, striped, tender" (Bridgeman).
Mertensia   Lungwort, Bluebells
  M. maritima. "Lungwort, elegant, sandy soil" (Breck).
  M. virginica. "Virginian Cowslip. Bluebells" (Breck).
  (Lady Skipwith's and Jefferson's "blue funnel flower.")
Mimulus   Monkey Flower
  M. cardinalis. "Brilliant scarlet flowers . . . varieties having rose or orange-
    colored blossoms" (Breck). "Hardy with winter covering" (Bridgeman).
  M. luteus. "Monkey Flower. Any soil or situation" (Buist, Bridgeman). "All
    shades of yellow . . . beautifully spotted or blotched with crimson or
    scarlet. It is a flower very much given to sporting" (Breck).
  M. moschatus. "Musk-scented. Small yellow flowers" (Breck, Buist).
    "Monkey-flower. Yellow, scarlet, rose" (Bridgeman).
  M. ringens. Allegheny Monkey flower (Breck).
Monarda   Horsemint, Beebalm
  M. didyma. "Oswego Tea. 3'. Scarlet" (Breck). "Scarlet ringlet flowers"
    (Buist). "Lemon-scented balm. Scarlet" (Downing). "Bergamot, crim-
    son, blue" (Bridgeman).
  M. fistulosa. "Light purple" (Breck, Bridgeman).
  M. punctata. Dotted monarda, horsemint. "Yellow and red" (Buist).
  M. russelliana (named for Dr. Russell who aided Nuttall in 1819 in Arkan-
    sas). "Red and white" (Buist).
Myosotis   Forget-Me-Not
  M. azorica. Azorean forget-me-not (Henderson).
  M. dissitiflora. Early forget-me-not. Form of m. sylvatica (Henderson).
  M. palustris (now M. scorpioides). True forget-me-not of Europe (Breck).

Nepeta sibirica (Breck)

Oenothera   Sundrops, Evening Primrose
  O. biennis. "Evening Primrose. Yellow" (Bridgeman).
  O. fraserii. "Fraser's Evening Primrose. Yellow" (Downing, Breck, Buist).
  O. fruticosa. Sundrops. "Shrubby, dark yellow" (Breck).
  O. macrocarpa (now *O. missouriensis*). "Showy, yellow" (Breck). "Large-podded Evening Primrose, yellow" (Downing, Buist, Henderson).
  O. pallida. (Buist).
  O. speciosa. "Large white flowers. Nuttall's white. Evening Primrose" (Breck, Buist).
Omphalodes   Venus's-Navelwort, Creeping Forget-Me-Not
  O. verna. "Brilliant blue. Common in English cottage gardens. Should be more popular here" (Breck). "Border flower" (Downing).

Paeonia   Peony
  Breck in 1866 declares that there are "one hundred varieties and species, all desirable." Breck begins with "the old double crimson peony (*P. officinalis*) familiar to everyone as a household friend—when first introduced into Antwerp about 1,500 plants sold for twelve crowns." He finishes his list with a handful of new varieties with fancy names from France and Germany, which we omit here as being no longer listed and only the first in a subsequent flood. It is obvious that there was confusion between herbaceous and tree peonies.
  P. albiflora (now *P. lactiflora*). From Siberia in 1756. "The white-flowered or Chinese Peony is the parent of many fine varieties" (Breck).
  P. festiva ('*Festiva maxima*'). "Double white with touches of purple" (Downing).
  P. moutan (now *P. suffruticosa*). Tree peony; var. *banksii,* light pink (Breck); var. *humeii,* double blush (Downing); var. *papaveracea,* Poppy-flowered, white, red at center (Downing); var. *rosea superba,* rose, double (Fessenden); var. *rubro-plena,* rose-colored, almost single. "Many tree peonies exhibited lately" (Breck). (Here again we have a list that is no longer recognizable. Breck mentions that his moutans do better with mulching in the winter and says several beautiful new specimens have been introduced by Mr. Fortune direct from China. Propagation is difficult from seeds, but Breck has raised some tree peonies from seeds from his own plants.)
  P. officinalis. "Flowering the last of May. Varieties include: *P. albicans,* white, *P. blanda,* blush; *P. rosea,* rose; *P. rubra,* red" (Breck).
  P. tenuifolia. "Fennel-leaved. Deep crimson. Early" (Breck, Downing, Fessenden).

Papaver Poppy

P. alpinum. "Alpine Poppy. White, Austria" (Breck).

P. bracteatum (now *P. orientale* var. *bracteatum*). "Bracted poppy, Siberia" (Breck, Downing, Bridgeman).

P. cambricum (now *Meconopsis cambrica*). Welsh poppy (Breck, Downing, Bridgeman).

P. nudicaule. Iceland poppy. "Siberia, yellow, red, 1 1/2'" (Breck).

P. orientale. "Most magnificent. Orange, scarlet" (Breck, Downing, Bridgeman, Fessenden).

P. pyrenaicum. "Yellow, red, 1'" (Breck).

Penstemon Beardtongue

P. barbatus. Bearded penstemon. Scarlet-orange flowers, half-hardy, splendid, from Mexico.

P. barbatus var. *torreyi.* "Torrey's Penstemon. Hardy species" (Fessenden).

P. campanulatus. Pale purple. Also known as *P. pulchellus* (Bridgeman, Downing, Breck).

P. cobaea. "Very showy, pale blue flowers, half hardy" (Breck).

P. coeruleus. "One of the finest of the genus. Beautiful blue flowers" (Breck).

P. digitalis. "Missouri Penstemon, White flowers" (Downing).

P. murrayanus. "Native of Texas, deep scarlet, half hardy" (Downing).

P. pubescens (now *P. hirsutus*). "Downy Penstemon. Beautiful pale blue flowers in panicles" (Breck, Downing).

P. richardsonii. "Purple-pink, hardy, from Oregon" (Breck, Downing).

P. speciosus. "Showy Penstemon. Flowers pale blue in loose panicles" (Breck, Downing). "Beautiful plants, peculiarly American, worth the attention of the amateur" (Breck).

Petunia

P. hybrida. "Many improved sorts" (Breck).

P. nyctaginiflora (now *P. axillaris*). "Large white flowers" (Breck).

P. violacea. "Purple Petunia. Introduced into England from South America 1831. Now indispensable" (Breck).

Phlox

P. divaricata (formerly *P. canadensis*). Wild sweet william. White, blue, purple, lilac. (Breck, Downing, Buist).

P. maculata. Wild sweet william. "Flora's Bouquet" (Breck). "Purple Spotted" (Downing).

P. paniculata. Summer phlox.

P. pilosa. "Hairy Phlox. Red. From which two all the most popular summer phloxes are descended." "Summer perennial phlox, purplish-red" (Breck, Bridgeman).

P. stolonifera, also *P. reptans.* Creeping phlox. "Dwarf. Creeping, red" (Breck).

P. subulata. "Moss Pink. Pink" (Breck). *P. subulata* var. *nivalis* (Bridgeman). Pink, purple, white, rosy-eyed (Buist, Downing, Fessenden).

Physostegia False Dragonhead

P. denticulata. "Lower, more slender" (Breck).

P. virginiana Obedience plant (Breck).

Platycodon grandiflorus Balloon Flower (sometimes listed under *Campanula grandiflora*)

Polemonium Jacob's Ladder

P. caeruleum. "Jacob's Ladder. Greek Valerian. Charity. One of the fine old border flowers. 2', June. Some white" (Breck). "Sweet-scented blue Valerian" (Bridgeman).

P. reptans. "Greek Valerian" (Downing, Fessenden).

Polygonatum Solomon's Seal

P. biflorum. Small (Breck, Bridgeman).

P. commutatum. Great (Downing, Henderson).

Potentilla Cinquefoil

P. astrosanguinea. "Dark blood red. Nepal, 1 1/2'. One of the finest" (Breck, Downing).

P. formosa splendens (now *P. nepalensis*). "Rose, puce, yellow" (Bridgeman).

P. grandiflora. "Large, yellow. Siberia" (Breck).

P. hopwoodii. "Splendid hybrid. Scarlet and rose" (Breck).

P. nepalensis, or *P. formosa.* Pink (Breck).

P. russelliana. "Splendid hybrid, scarlet and rose" (Breck, Buist, Downing).

Primula

P. chinensis (now *P. sinensis*). "Chinese Primrose. Lilac, white" (Bridgeman).

P. cortusoides (now *P. sieboldii*). "Hardy" (Fessenden, Buist).

P. elatior. Oxlip (Buist).

P. farinosa. Bird's-eye. "Small" (Buist).

P. polyanthus. "Commonest hardy primula" (Breck). "The Polyanthus. Purple" (Downing).

P. scotia. "Small. Half as large as *P. farinosa*" (Buist).

P. suaveolens–P. veris var. *suaveolens.* (Buist).

P. veris. Cowslip. "Native of Britain. Mostly yellow, some red, some double" (Breck). "Cool, shady" (Fessenden).

Pulmonaria Lungwort, Jerusalem Sage

P. officinalis. "Red and bluish, spotted leaves" (Breck).

P. saccharata. Bethlehem sage.

P. virginica (see *Mertensia*).

Pyrethrum
   P. carneum (now *Chrysanthemum coccineum*). "Rosy Pyrethrum. New double
   varieties" (Breck).

Ranunculus   Buttercup, Crowfoot
   R. aconitifolius 'Flore pleno.' "Fair Maids of France, double, white, beautiful,
   1'" (Breck).
   R. acris flore pleno. "Bright yellow, upright, double, 2'. Tall Buttercup"
   (Breck).
   R. repens flore pleno. "Troublesome because creeping" (Breck).
   R. asiaticus. "Persian Buttercup" (Breck).
Rudbeckia   Coneflower
   R. fulgida. "Bright yellow, dark centre, July, August" (Breck).
   R. lutea. "Yellow" (Bridgeman).
   R. nudiflora. "Conical dark disks, yellow flowers" (Breck).
   R. purpurea (now *Echinacea purpurea*). "Purple Rudbeckia. 3'–4', rich"
   (Breck, Bridgeman). "North American plants, hardy, some borders"
   (Breck). "Many flowers . . . very durable . . . and much admired"
   (Fessenden).

Sabbatia
   S. stellaris. Sea pink. Named by Pursh. "American Centaury. Large showy
   pink flowers in July" (Breck).
Sanguinaria canadensis
   "Bloodroot. Red Puccoon. The Indians are said to paint their faces with the
   juice" (Breck).
Saponaria   Soapwort
   S. ocymoides. "Basil-like Soapwort, red" (Downing).
   S. officinalis. Bouncing Bet. "Good in garden" (Buist).
   S. officinalis plena. Double soapwort. "Good" (Bridgeman).
Sarracenia
   S. purpurea. Pitcher plant. "Side-saddle flower. Pitcher plant. Cultivated be-
   fore 1640 by Tradescant" (Breck).
Saxifraga
   S. crassifolia (now *Bergenia crassifolia*). "Heart leaved saxifrage. Pink pan-
   icles. May–June. 1'" (Breck). "Thick leaved saxifrage. Lilac" (Downing,
   Bridgeman). "Used for tanning" (Buist).
   S. granulata flore pleno. Fair maids of France. "Double white, desirable"
   (Buist).

S. umbrosa. "Saxifrage. Rose, white, purple" (Bridgeman). "London-pride, good edging. None so pretty" (Buist).

S. virginiensis. "Fragrant. Earliest flowers upon rocks and dry hills" (Breck).

Sedum   Stonecrop, Live-Forever

S. acre. "Stonecrop" gold moss (Henderson).

S. populifolium. "Subshrub. Poplar-leaved sedum, white" (Downing).

S. pulchellum. "Beautiful stonecrop." (Henderson).

S. sieboldii. "Siebold's Stonecrop. Considerable beauty and interest" (Breck).

S. spectabile. "Showy Stonecrop" (Henderson).

Sempervivum tectorum   Common Houseleek

"Of these curious plants there are more than fifty species in cultivation, and all perfectly hardy; useful on rock work" (Henderson).

Senecio   Groundsel.

S. aureus. "Golden Ragwort." "Golden Senecio. Handsome. Indigenous. Not often introduced into the flower border although much handsomer than many plants that are cultivated" (Breck).

Solidago   Goldenrod.

"A few look pretty in gardens" (Breck).

S. bicolor. Silverrod (Breck).

S. odora. "Sweet-scented Golden Rod. Pleasant . . . odor" (Breck).

Spiderwort (see *Tradescantia*)

Spiraea

S. aruncus (now *Aruncus sylvester*). "Goat's beard. 3'–4', white" (Breck).

S. filipendula (now *Filipendula hexapetala*). "Dropwort. Beautiful foliage, pink buds white flowers. June, July" (Breck).

S. japonica. "More delicate than *S. aruncus*" (Breck).

S. palmata (now *Filipendula palmata*). "Magnificent, from West, 5'–6'" (Breck).

S. ulmaria plena (now *Filipendula ulmaria plena*). "Double Meadow Sweet. Handsome border plant. White, 2'" (Breck).

Statice (now rejected as a botanical name; the thrifts are now *Armeria*; Sea lavender is *Limonium*)

S. arborea. "Now [1838] selling in England for Fifty dollars each" (Buist).

S. latifolium and *S. maritimum.* "Finest" (Buist).

S. speciosum (now *Goniolimon speciosum*). "Among prettiest" (Breck). "Red" (Bridgeman).

S. tataricum (now *Goniolimon tataricum*). "Among prettiest" (Buist). "Among prettiest and hardiest" (Breck).

S. vulgare. "Thrift, red, white." (Bridgeman). "Shear when through flowering." (Buist).

Thalictrum Meadow Rue

   T. anemonoides (now *Anemonella thalictroides*). "Rue anemone. Common plant in woods . . . white sometimes tinged with pink" (Breck).

Tiarella Foam Flower, False Mitrewort

   T. cordifolia. "Heart-leaved Tiarella. White" (Breck).

Tradescantia Spiderwort, Widow's Tears

   T. virginiana. "Valuable border flower May–September. Blue, white, reddish purple" (Breck). "Blue and white Spiderwort" (Downing).

Trollius Globeflower

   T. asiaticus. "Asiatic Globe Flower. Yellow" (Bridgeman). "Large dark orange flowers. 1′ June, July" (Breck).

   T. europaeus. "Globe Crow-foot. Bright yellow" (Breck). "European Globe Flower. Yellow" (Downing).

Uvularia Bellwort, Merrybells

   U. grandiflora. "Large flowered Bell-wort. Flowers yellow, worth cultivating" (Breck).

   U. perfoliata. Strawbell. "Perfoliate Bell-wort. Pale yellow, May" (Breck).

Valeriana Valerian

   V. officinalis. Garden heliotrope. "Garden Valerian. Small white fragrant flowers. 3′–4′" (Breck).

   V. pyrenaicum. "Heart-leaved, pink, moist soil" (Breck).

   V. rubra v. *alba* (now *Centranthus ruber*). Red valerian. Jupiter's beard. "Garden Valerian, red, white" (Bridgeman).

Verbascum

   V. phoeniceum. "Purple Mullein" (Downing).

   V. thapsus. Common Mullein.

Verbena canadensis

   "A weedy race until the introduction of . . . *V. lambertii*" (Breck).

Veronica Speedwell

   V. azurea. "Sky blue. 2′–3′" (Breck).

   V. coerulea. Blue (Bridgeman).

   V. gentianoides. "Gentian-leaved Speedwell, blue" (Downing).

   V. maritima (now *V. longifolia*). "Maritime Speedwell, blue" (Downing).

   V. sibirica. "Blue spikes, July, August" (Breck).

   V. speciosa (now *Hebe speciosa*). "Brilliant blue, dwarf" (Breck).

   V. spicata. "Fine blue, 1′" (Breck). "Blue spiked Speedwell" (Downing).

   V. variegata.

   V. virginiana. "White, July, August, 4′–5′" (Breck).

(Buist lists a whole catalogue at the end of his book: *V. officinalis, chamaedrys, media, incana, spicata, grandis, incarnata, carnea, leucantha, bellidioides, verna, amoena, pulchella.*)

Veronicastrum virginicum   Culver's Root, Bowman's Root

Vinca   Periwinkle, Myrtle

  V. alba, V. rosea. "Periwinkle, Madagascar. White, rose" (Bridgeman).

  V. major. Blue-buttons (Breck).

  V. minor. "Myrtle, evergreen. Blue flowers" (Fessenden). "Several species, all neat and pretty in their place" (Breck).

Viola   Violet

  V. alba "Violet, fragrant, white." (Bridgeman).

  V. odorata. "Sweet scented violet" (Breck). "Violet, blue, fragrant. Of considerable use in chemical inquiries to detect an acid or an alkali; the former changing the blue color to a red and the latter to a green" (Fessenden). "Fragrant, blue" (Bridgeman).

  V. odorata plena. "Double white and blue European violets" (Downing).

  V. tricolor. "Pansy. Lady's Delight. Heart's Ease" (Breck). "Many colors and sorts" (Downing).

Viscaria   German Catchfly

  V. vulgaris (now *Lychnis viscaria*). "White and red Viscaria" (Downing).

VI   *Roses*

Roses in mid-nineteenth-century gardens luxuriated in named hybrids. The first hybridization took place in Charleston, South Carolina, in the garden of a rice planter, John Champneys, at the begininng of the nineteenth century. By then there were four China roses being grown in England and Bermuda, and as near as Charleston, where they still flourish miraculously. They were Slater's Crimson, Parson's Pink, Hume's Blush, and Park's Yellow, all everblooming. The old European garden roses bloomed only once. Champneys succeeded in crossing the old musk rose (*R. moschata*) with Parson's Pink China and called his result Champney's Pink Cluster. One of the Noisette brothers, eminent French nurserymen, was in Charleston and sent the rose to his brother in Paris, who produced from it the whole race of Noisette roses. At the same time, a Parson's Pink China, growing on the Isle de Bourbon (now Réunion) crossed with the only other rose on the island, a pink Autumn Damask, and started the race of Bourbon roses. Also, in the early nineteenth century the empress Josephine at Malmaison was assembling the best collection of roses in the world. No wonder the excitement of new sorts of roses overpowered gardeners who had been quite happy with what we now call the old shrub roses, the gallicas, damasks, albas, spinosissimas, mosses, and eglantines. Single roses had to give way to enormous heavy blooms, the bigger the better. Luckily, they were all still fragrant, although fragrance was gradually to be bred out in the striving for size and new colors.

After giving M'Mahon's basic list of 1806, we are following here chiefly the lists of roses grown for sale by the foremost breeder of his time, Robert Buist of Philadelphia, as described in his *American Flower Garden Directory,* 1839, and in the lists recommended by Downing in his *Cottage Residences,* 1847, and by Francis Parkman in *The Book of Roses,* 1886.

M'Mahon in 1806 listed in his catalogue of "Hardy Deciduous Trees and Shrubs" roses that included all the old favorite European rose species with varieties and several American natives. The few Chinese roses are all considered "tender."

Rosa berberifolia   Single-leaved Rose
Rosa lutea   Single Yellow Austrian Rose
Rosa sulphurea   Double Yellow
Rosa blanda   Hudson's Bay Rose
Rosa cinnamomea   Cinnamon Rose
Rosa arvensis   White Dog Rose
Rosa pimpinellifolia   Small Burnet-leaved Rose
Rosa spinosissima   Scotch Rose—with four varieties, red, striped, double, common
Rosa parviflora   Small flowered—American or Pennsylvania Rose
Rosa lucida   Shining-leaved American Rose
Rosa carolina   Carolina Rose
Rosa villosa   Apple Rose
Rosa rugosa   Wrinkled-leaved Rose
Rosa provincialis   Common Province Rose with six varieties: Childing's, Red, Blush, White, two de Meaux-major and minor, or Pompons
Rosa centifolia   Hundred-leaved Rose with eleven varieties: Dutch, Blush, Singleton's, Burgundy, Single Velvet, Double Velvet, Sultan, Stepney, Garnet, Bishop, Lisbon
Rosa gallica   Red—official Rose with three varieties: *versicolor* Mundi, Marble, Virgin
Rosa damascena   Red Damask Rose with seven varieties: Blush, *versicolor* York and Lancaster, Red Monthly, White Monthly, Blush Belgic, Great Royal, Imperial Damask
Rosa pumila   Dwarf Austrian rose
Rosa turbinata   Frankfort Rose

Rosa rubiginosa   Sweet Briar with seven varieties: Common, Double, Mossy Double, Marble Double, Red Double, Royal, Yellow
Rosa muscosa   Moss Province Rose
Rosa moschata   Single Musk Rose with a double variety
Rosa rubrifolia   Red-leaved Rose
Rosa lagenaria   Bottle-fruited Rose
Rosa alpina   Alpine Rose
Rosa pyrenaica   Pyrenean Rose
Rosa pendulina   Pendulous Rose
Rosa montana   Mountain Rose
Rosa multiflora   Many-flowered Rose
Rosa canina   Dog Rose or Dog Briar
Rosa tomentosa   Downy-leaved Rose
Rosa collina   Hill Rose
Rosa parvifolia   Small-leaved Rose
Rosa longifolia   Long-leaved Rose
Rosa alba var. *pleno*   Double White Rose, with four varieties: Small Maiden's Blush, Large and Small ditto, Cluster, Small burnet ditto
Rosa semperflorens (tender)   Everblowing Rose
Rosa chinensis (tender)   Pale Chinese or Otaheite Rose
Rosa indica (tender)   Indian Rose
Rosa palustris   Swamp Rose

By 1839, in his *American Flower Garden Directory*, Robert Buist, who had taken over much of M'Mahon's nursery grounds, was able to recommend a list of 250 roses. He presents these as "the finest," although he estimates there were then at least 2,000 varieties of the garden rose. He breaks his presentation into two parts: roses to be set out in March, and those more tender, to be set out in April. Of his seventy or more March set-outs, we note several of today's standbys: Belle Hebe, Black Tuscany, Cabbage Provins, Harrison's, Madame Hardy, White Bath Moss.

His April set-outs are broken down into classes for us:

1. "Hybrid Chinese," of which he lists twenty-seven, none easily obtainable today

2. "Hybrid Roses that are Striped, Spotted or Marbled," of which he lists eight including today's Camaieux

3. "Perpetual Roses," which he says prove hardy only when grafted onto our strong-growing native roses, and where, among his sixteen entries, we find our Rose du Roi or "Lee's Crimson Perpetual, admittedly the king of perpetuals"

4. "Isle de Bourbon Roses," seven "cheerfully recommended," among which we know Madame Desprez

5. "Rose indica or Bengal (of the French) Chinese Everblooming Roses," of whose thirty-six we grow Cramoisie superieur

6. "Rosa odorata or Tea Rose," with twenty-four entries among which Madame Desprez "pure white, universal favourite" appears as "Bengal Madame Desprez"

7. "Noisette Roses," with twenty-five entries, among which is Agrippina, which grows so beautifully today in Bermuda, and Champney's Pink Cluster, also surviving there

8. "Rosa moschata," with only four, of which he has two double

9. "Climbing Roses that Bloom Only Once in a Season," with nineteen, of which many are Boursault (which he spells "Boursalt" and says they all withstand severest cold) and Noisette. Banksia roses, both white and yellow are here, apparently much as they arrived from China to take over southern gardens. Another Chinese import is the Seven Sisters or Grevillia, which survives in New England. *Multiflora* white and pink ("first climbing rose in Philadelphia, twenty dollars for a single plant")

10. "Rosa microphylla or Small-leaved Rose," with four entries, all with flowers in twos and threes, dark toward the center, exterior petals large and full, interior short and thick with white, pink, and yellowish blooms

Because this is our best opportunity to bow to southern readers who have long reveled in Chinese imports like the Cherokee rose (R. *laevigata*) and Macartney (R. *bracteata*), long cultivated the R. *microphylla,* and cherished the green rose (R. *chinensis viridiflora*), to say nothing of Mr. Champneys's success with crossing Parson's Pink China, we can only wonder at those early traders seemingly unheralded today who introduced these roses as well as the Oriental camellias and forsythias and quinces apparently long before they were formally introduced to the world by London botanists and gardeners. Someday, someone—

Our American climate range being what it is, however, it may be better for our mid-nineteenth-century garden lists to cease to wonder and instead envy the present simple working list that A. J. Downing suggests in his *Cottage Residences,* the third edition, of 1847. He gives us short lists of the "beautiful monthly roses":

> *Bourbon*
> Souvenir de la Malmaison, large, shell color
> Paul Joseph, rich deep crimson

Madame Angelina, white, fringed with fawn
Souchet, rose tinted with carmine
Acidalie, white or pale blush
Madame Desprez, rose, large and very double
General Dubourg, ditto, very fragrant
Hermosa, rose-colored
Gloire de France, or Neumann
Queen, rosy fawn

Bengal
Mrs. Bosanquet, pale flesh
Louise Phillippe, deep crimson
Agrippina, crimson or striped
Double White Daily
Queen of Lombardy, cherry color

Noisettes
Champney's Cluster, blush
Aimée Vibert, pure white
Fellemberg, crimson
Cloth of Gold, fine yellow
Conque de Venus, delicate blush
Jaune Desprez, creamy blush
Lamargue, pale yellow
Smith's Yellow, large and fragrant
Grandiflora, large blush
Sir Walter Scott, purple
Joan of Arc, white
Charles X, bright red

Tea Roses
Odorata, or Common Blush Tea
Devoniensis, creamy white
Caroline, fine blush
Josephine Malton, yellowish white
Bourbon White Tea
Semperflorens, or Sanguinea
Roi des Cramoisies, rich crimson
Marjolin, superb dark red
Leonidas, bright rose
Bougere, glossy fawn
Aurora, pale straw

Clara Sylvian, fine
Goubault, bright rose

Francis Parkman, our eminent American historian, was also a grower of lilies and roses. In his *Book of Roses* of 1866 we have a whole new wave of roses, especially in the category of "Hybrid Perpetuals." Under the heading "Most Approved by the Best Cultivators of the Present Day," he adds 600 roses to those he has already given us. "Cultivated Roses Once Blooming" are listed in divisions of: Provence, Moss, French, Hybrid China, Damask, Alba, Austrian Brier, Sweetbrier, Scotch, Double Yellow (*sulphurea*), Ayrshire, Semper-virens, Multiflora, Boursault, Banksia, and Prairie. Under "Perpetual or Re-montant" he lists: Tea, Bourbon, Hybrid Perpetual, Perpetual Moss, Damask Perpetual, Noisette, Musk, Macartney, and Microphylla. Most of these are our old friends except for the introduction of the strong yellow and orange colors of the Austrian yellow and copper, and the new varieties of the little Scotch rose, *R. spinosissima*. Parkman hails the old Provence rose, *R. centifolia* or Cabbage Rose, as "remarkable among roses" for its sports like the crested Moss, striped Unique, and miniatures like De Meaux.

It is worth noting that over two-thirds of his "new" list is devoted to the hybrid perpetuals, predictably mostly French-named.

And with this the new world of rose breeding is under way.

Walter Elder's contribution (in *The Cottage Garden of America*) to the records of rose growing in cottage gardens in midcentury lies in his account of costs, though the differences in prices are left for us to attribute to lack of popularity or quantities available.

Rosa indica, "The Daily Rose," includes: Agrippina, Cels, Grenadier, Hor-tencia, Indica, Indica Alba, Louis Philippe, Lady Warrender, Mrs. Bosanquet, Napoleon, Prince Eugene, and Reine de Lombardie, to be bought for three to five dollars a dozen.

Rosa Indica odorata, "The Teascented rose," includes: Aurora, Caroline, De-voniensis, Goubault, Gloire de Hardy, La Sylphide, Madame Desprez, Man-sais, Odoratissima, Triomphe de Luxembourg, and William Wallace, all five to nine dollars a dozen.

The Bourbon Roses include: Acidalie, Doctor Roques, General Dubourg, Gloire de Rosamanes, Hermosa, Henry Clay, Henri Plantier, Le Phoenix, Madame Desprez, Paul Joseph, Queen, and Souvenir de la Malmaison, all for six dollars a dozen.

Noisette Roses, for which M. Noisette received all credit, include: Aimée Vibert, Camellia Rouge, Cloth of Gold, Champney, Fellenberg, Jane, Jaune Desprez, Lee Mostroia, Ophire, Sir Walter Scott, Solfaterre, and Sultana, all four to eight dollars a dozen. They are tender in New York City.

Hybrid Perpetuals include: Comte de Paris, Emma Dampierre, Lady Ford-wich, Louis Bonaparte, Lady Alice Peel, Madame Laffay, Prudence Roeser, Princess Helena, Prince Albert, Rivers, Reine de la Guiliotiere, and Rose de la Reine, twelve dollars a dozen.

Musk-Scented Roses include: Moschate, Princesse de Xlassau, Herbemont's, and Red Musk Cluster at thirty to fifty cents each.

Perpetual Damask Roses include: Claire de Chatelet, Du Roi or Lee's, Jenny Audio, La Reine, Palmyra, and Monthly Damask, nine dollars a dozen.

Microphylla Roses include: Maria Lonida, Rosea, and Purpurea, thirty-five to fifty cents each.

Climbing Roses Flowering Once a Year include: (1) Boursault—Alba, Purpurea, Inermis—thirty to fifty cents each; (2) Multiflora—Alba, Lawrie Davoust, Cottage, Grevillei—thirty-five to fifty cents each; and (3) Prairie Rose—Baltimore Belle, Queen of the Prairies, Superba, Pallida, Michigan, and Kentucky—thirty to fifty cents each.

Hardy Bush Roses that Bloom Once a Year include: Moss, Cabbage, Rose of France, Yellow, Marbled, White Garden, Damask, and Hybrid Chinese, three to six dollars a dozen.

VII   *Shrubs*

Shrubs, which often included small trees, were sought after to plant in groups with other shrubs of varied foliage and flowering habit, to stand alone as accents in the landscape, or to fill in lower levels of garden grounds to make smooth contours, as was done in the last part of the eighteenth century by landholders like Washington and Jefferson. With the nineteenth century and the rise in garden elegance of the small landholder and cottager, shrubs were fancied for individual features of bloom and fruit and autumn color. They were used to form shields against neighbors—though not against the street and the passing stranger. They became fillers to awkward corners, or individual features set about on the lawns. They could also be clipped into different shapes to bloom in balls or squares or when espaliered against walls.

Aesculus parviflora   Bottlebrush Bush
   "Too well known to need description" (Henderson).

Amorpha fruticosa   Bastard Indigo
   "Native to South Carolina. Once used as an indigo plant . . . spikes of purplish flowers in July" (Breck). "Very desirable, prefers shade" (Buist).

Amygdalus pumila plena (now *Prunus amygdalus*)   Double Dwarf Almond
   "Elegant flowers resembling small roses." "Most common early spring shrub" (Scott).

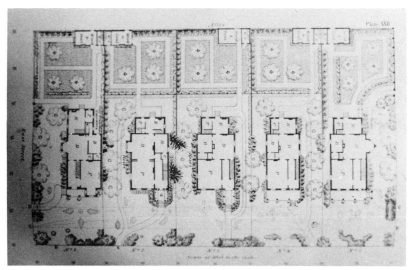

Suburban plot plan. From Scott, *Beautifying Home Grounds*.

Andromeda   Bog Rosemary
"A great number of North American species might be introduced into the shrubbery with good effect" (Breck). Buist recommends "all the species."
A. floribunda (now *Pieris floribunda*). "An evergreen shrub" (Henderson).
A. polifolia. Water andromeda. "Flowers are red before they open" (Breck).
A. speciosa (now *Zenobia pulverulenta*). "Flowers in great profusion" (Breck).
Aucuba japonica   Japanese Laurel
"Very desirable, prefers shade" (Buist).
A. japonica var. *variegata* (now *Pieris japonica*). Gold-dust tree. "Evergreen. Handsome foliage with golden spots" (Breck).
Azalea (now all classed under *Rhododendron*)
A. arborescens (now *Rhododendron arborescens*). Treelike azalea. "The finest ornamental shrub Pursh knows. Rose colored" (Scott).
A. nudiflora (now *Rhododendron nudiflorum*). "Occasionally presents itself to the enraptured traveler, tempting him for a while to forget the objects of his journey, and admire the elegance and fragrance of its flowers" (Breck). "Upright American Honeysuckle. Flowers pink, white, striped, red, and purple. Superseded in cultivation by new varieties" (Scott).
A. viscosa (now *Rhododendron viscosum*). "Called Swamp Pink by the country people probably on account of the odor of the flowers" (Breck). "Flowers

leafy, hairy, and sweet-scented, white" (Scott). In 1888 Henderson mentions *A. pontica* hybrids as "Belgian Azaleas, rose, yellow, etc."

Benzoin aestivale (see *Lindera benzoin*)
Berberis   Barberry
  B. dulcis (now *B. buxifolia*). "Dwarf foliage, delicate and almost evergreen. Makes a handsome hedge" (Breck).
  B. vulgaris. Common barberry. "Arching upper shoots . . . sustaining racemes of splendid yellow flowers . . . succeeded by clusters of scarlet fruit" (Breck). "Grown in England for its fruit" (Scott). Buist recommends "all the species."
Buxus   Box
  B. sempervirens. "Garden Box—A delicate shrub which may be pruned to any shape to tease the fancy . . . in general use, best material for forming edgings to beds, walks, etc. Plants may be trained singly into almost any shape, and will make large shrubs . . . varieties with yellow and white striped leaves, called gold and silver . . . a number of species, among which are dwarf and tree box, the last suitable for the shrubbery" (Breck).
  B. suffruticosa. Edging box. "Dwarf box—Planted less than it deserves to be" (Scott).

Cerasus   Sour Cherry, Pie Cherry
  C. communis plena (now *Prunus cerasus*). "Double-flowering cherry. Desirable addition to the shrubbery . . . flowers are like small white roses" (Breck).
Chionanthus virginicus   Fringe Tree
  "Most elegant little tree that can grace a lawn" (Scott).
Clematis recta
  White, sweet-scented. Herbaceous.
Clethra alnifolia   Sweet Pepperbush
  "Long spike of white fragrant flowers toward the end of summer" (Breck). "Though indigenous to our woods, it has been brought into notice in the New York Central Park" (Scott).
Colutea arborescens   Bladder Senna
  "Yellow or orange pea-shaped flowers . . . followed by seed-vessels like bladders" (Breck). "A good shrub for the interior of masses of shrubbery" (Scott).
Cornus   Dogwood
  C. alternifolia. "Recent shoots of a shining light yellowish-green" (Breck).
  C. circinata (now *C. rugosa*). "Round-leaved, blotched with purple" (Breck). "Noted for large circular wavy leaves" (Scott).
  C. mas. Cornelian cherry.

C. paniculata (now *C. racemosa*). "Pale yellow stems, flowers white . . . white fruit . . . when fruit stalk is of a delicate pale scarlet" (Breck). "One of the best for small grounds" (Scott).

C. stolonifera (now *C. sericea*). "Blood-colored stems and shoots" (Breck). "There are few more pleasing shrubs in the Central Park" (Scott).

Cotinus coggygria Smoke Tree, Venetian Sumac (see *Rhus cotinus*)

Cotoneaster

"Mostly half-hardy" (Scott). *C. microphyllus* is recommended by Henderson. Buist says "all are fine."

Cydonia japonica (now *Chaenomeles japonica*) Japan Quince

"The hue of the scarlet color is most brilliant, and no artist can find a tint that will convey an adequate idea of its splendor" (Breck). "Too well known to need description" (Scott).

Cytisus scoparius Common Broom, Scotch Broom

"Very ornamental shrub in interior of country" (Breck).

Daphne Evergreen

"Of value" (Scott).

D. cneorum. "Garland Flower" (Henderson).

D. mezereum. "Flowers come out before the leaves. Brilliant scarlet berries said to be a powerful poison. Another variety with white flowers, has yellow berries. . . . Blossoms will perfume the air to a considerable distance" (Breck). "Valued for the earliness of its bright blossoms" (Scott).

Deutzia

D. crenata. "The variety most largely disseminated. The double white and the double pink are the finest large sorts and should be planted near together where the colors will be contrasted" (Scott).

D. gracilis. "Native of Japan and China. Related to *Philadelphus*. Introduced from China a few years ago by Mr. Reeves, to whom our gardens are indebted for many other equally interesting plants from the same quarter. Covered by a profusion of white blossoms, highly fragrant. The rough leaves of the plant, Thunberg informs us, are employed by the Japanese cabinet-makers for polishing wood, in the manner that the stems of *Equisetum hyemale* are with us" (Breck). "Smallest variety and greatest favorite" (Scott).

Diervilla trifida (now *D. lonicera*)

"Three-flowered Bush Honeysuckle—Three yellow honeysuckle-shaped flowers at each leaf axil" (Breck).

Dirca palustris Moosewood, Leatherwood

"Yellow blossoms. Found in wet, marshy and shady places. Wood very pliable, and the bark . . . has such strength that a man can not pull apart so much as covers a branch of half an inch in diameter . . . used by millers and others for thongs. The aborigines used it as a cordage" (Breck).

Elaeagnus

E. *angustifolia* (formerly E. *hortensis*). Oleaster. Russian olive. "Old English garden shrub . . . for the silvery whiteness of its foliage . . . is often selected to plant where it is desired to attract attention or to create variety" (Scott).

E. *argentea* (now E. *commutata*). Silverberry. "Elegant silvery foliage their only recommendation" (Breck). "Missouri Silver Tree—Fine specimen growing near Seventh Avenue entrance to Central Park" (Scott).

E. *japonica* (now E. *parvifolia*). "Small-flowered shrub noted for whitish foliage" (Scott).

Erica Heath

Was considered a house plant.

Euonymus Spindle Tree

E. *americana*. Strawberry bush. Bursting heart. "Inconspicuous purple flowers in clusters . . . succeeded by brilliant scarlet fruit" (Breck). "Pretty little umbrella-shaped tree . . . planted for its brilliant and curious seed vessels" (Scott).

E. *atropurpurea*. Burning bush, Wahoo. "A native variety distinguished by its purple flowers" (Scott).

E. *europaea*. "Smaller than our own and not so handsome" (Scott).

E. *latifolia*. "Most beautiful foliage of the family . . . but not so brilliant in color as the American" (Scott).

E. *radicans*. "New variety from Japan . . . of striking beauty" (Scott). "Euonymus is popularly known by the names Strawberry Tree, Spindle Tree, and Burning Bush. Burning Bush so called because of the pendant scarlet seed capsules" (Scott). (Rehder calls E. *americana* "Strawberry Bush" and E. *atropurpurea* "Burning Bush.")

Exochorda Pearl Bush

E. *grandiflora* (now E. *racemosa*). Large flowering Spirea—white (Henderson).

Forsythia viridissima Golden Bell

"Introduced from the North of China in 1845. Earliest of spring flowering shrubs" (Henderson). "Large spreading shrub . . . of brilliant green foliage. A comparatively new thing . . . a reputation it may not sustain. Should be grown among other shrubs" (Scott).

Fothergilla alnifolia (now *Fothergilla gardenii*) Dwarf Alder, Witch Alder

"Spreading native shrub. Flowers small, sweet-scented, appear before the leaves" (Scott).

Halesia Snowdrop, Silver-Bell Tree

H. carolina (formerly *H. tetraptera*). "Four-winged Halesia—Native of Carolina. Snowy-white flowers hang in small bunches all along the branches" (Breck). "Low spreading . . . a profusion of pure white, pendant flowers resembling those of the snowdrop" (Scott).

H. diptera. Two-winged silver-bell tree. "It differs very strongly from the common species, in both the larger size and the purer white of the flowers, and also in foliage, which is twice as broad as that of the four-winged sorts" (Downing).

Hibiscus Rose Mallow

H. syriacus. Rose of Sharon. "Althaea frutex—Single varieties are equally handsome as the double, and generally more hardy. There is the double white, red, blue, with stripes or blotches, and others" (Breck). "One of the longest known and commonest of the garden shrubs . . . forms a good center for a group of lower shrubs" (Scott).

H. syriacus flore pleno. "Rose of Sharon—Double, all shades from white to crimson" (Henderson).

Hydrangea

H. cordata and *H. arborescens.* "Natives of Pennsylvania and Virginia . . . *H. canescens* is a low shrub of the southern states, flowers larger than the preceding" (Scott).

H. hortensis (now *H. opuloides*). "Seen in or near almost every New England village porch. Most beautiful outdoor-box plant" (Scott). "Common hydrangea blue or pink. Sent from China by Sir Joseph Banks in 1790. Since then many have been sent from Japan. White variety now very popular named 'Thomas Hogg'" (Henderson).

H. paniculata var. *grandiflora.* "Pee Gee." "Japan Hydrangea—Recently introduced . . . bids fair to be the most valuable of the hydrangeas . . . hardy in the Central Park" (Scott).

H. quercifolia. "Hardier than *hortensis*. Sheltered situation, moist soil. Leaves turn a fine deep red color" (Scott).

Hypericum

H. calycinum. Rose of Sharon. Aaron's beard. "Evergreen trailing with larger leaves and flowers . . . greatly esteemed . . . for planting among rocks and trees" (Scott).

H. kalmianum (evergreen) and *H. prolificum* (shrubby broombrush). "Broad, compact, low shrubs . . . corymbs of small yellow flowers . . . highly valued" (Scott).

Ilex

    I. aquifolium. English holly. "Evergreen. Highly esteemed for the shrub-
bery. Silver and gold varieties are very beautiful. The common green
prickly-leaved is used for hedges; the only objection to it is its very slow
growth. . . . They may thrive in sheltered places where not much exposed
to the sun. They are worthy of many trials" (Breck).

    I. opaca. American holly. "In the North, half-hardy shrub" (Scott).

Jasminum

    J. nudiflorum. "Recently introduced from China. Yellow blossoms. One of
the prettiest shrubby bushes imaginable. Requires protection" (Scott).

    J. officinale. Poet's jessamine. "Really a noble climber in congenial climates
. . . may be grown as a shrub. Requires winter protection in the north.
White fragrant blossoms" (Scott).

Juniperus

    J. canadensis (now *J. communis depressa*). "Low spreading shrubs that take up
a great deal of room that may be much more prettily occupied by other
things" (Scott).

    J. prostrata (now *J. horizontalis*). "Fine specimens in the Central Park"
(Scott).

Kalmia

    "Evergreen shrubs known as the American or mountain laurel. Not been
much used for embellishment" (Scott).

    K. angustifolia. Sheep laurel. "Red flowers" (Breck). Breck quotes Cobbett's
caustic comment regarding *K. angustifolia:* "The little dwarf brush stuff,
that infests the plains of Long Island, is, under a fine Latin name, a choice
greenhouse plant in England, selling for a dollar when no bigger than a
handful of thyme."

    K. glauca (now *K. polifolia*). "Red flowers"

    K. latifolia. Calico bush. American laurel. "Thick glossy leaves, flowers
white to red." "Charming where growing in soil, shade and moisture"
(Scott).

    "Mountain Laurel, Spoon Wood, etc.—There is no shrub, foreign or native,
that will exceed this in splendor, when well grown" (Breck).

Kerria japonica

    "Low shrub, yellow flowers, sometimes all summer" (Scott). "Japan Globe
Flower—An old favorite in the garden with single and double flowers"
(Henderson).

Laburnum anagyroides   Golden Chain

"Laden with long pendulous clusters of golden pea-shaped flowers" (Breck).

Laurus benzoin (see *Lindera benzoin*)

Lavandula angustifolia (formerly *L. spica*). "Spike-flowered lavender—Named for its use in fomentations and baths. Desirable dwarf shrub. The whole plant is delightfully fragrant, but more particularly the flowers. . . . It is sometimes used for edging in milder climates, but grows too high for general use. As an edging for a bed of Moss Roses, we have seen it used with pleasing effect" (Breck).

Ligustrum vulgare   Common Privet, Prim

"Deciduous. White blossoms. The blossom of the Privet, when exposed to the noonday sun, withers almost as soon as blown. In the shade, it not only lasts longer, but is much larger and finer when so placed" (Breck). "Numerous varieties—no shrub bears clipping better" (Scott).

Lindera benzoin   Spice Bush, Benjamin Bush

"Fever Bush—Spice Bush—The plant derives its botanical name from its aromatic odor, resembling gum benzoin" (Breck).

Lonicera tatarica   Tatarian Honeysuckle

"Native of Russia. Pink flowers followed by red berries. There is a variety with white flowers and yellow berries" (Breck). "Red Tatarian Honeysuckle—Old and common, it still takes the front rank among ornamental shrubs. . . . were we to have but one shrub, we would probably choose the honeysuckle" (Scott).

Maclura pomifera   Osage Orange

"The favorite hedge plant of the United States . . . planted by gentlemen of wealth and taste around their favorite walks and grounds when the plants sold at the rate of five dollars per thousand. . . . Hedges of the rarest beauty and excellence have been growing in Boston, Philadelphia, and Cincinnati, in Kentucky, Tennessee and northern Missouri" (Porcher, 1848, in the *Prairie Farmer*).

Magnolia

M. *acuminata*. Cucumber tree. "If Dr. Kirtland's success in growing this variety of the *M. acuminata* can become general, we shall have one of the best ornaments of our lawns. On its own roots . . . a scrawny bush" (Scott).

M. *glauca* (now *M. virginiana*). Sweet bay. "Few ornamental plants are better worth the attention of the gardener. . . . The flower, pure white, two or three inches broad, is as beautiful and almost as fragrant as the White Lily" (Emerson). "The Swamp Magnolia is rather a large bush than a tree. Blossoms snowy white and so fragrant that where they abound

in swamps, the perfume is often perceptable for a quarter of a mile"
(Downing).

Mahonia (named for Bernard M'Mahon)   Oregon Grape

   M. aquifolium. "Ilex-leaved Mahonia—Evergreen with clusters of yellow
   flowers, followed by blue berries. The foliage is very gay, as on the same
   plant there will be bright-green, purple, brown and crimson leaves"
   (Breck). "Mostly natives of the vallies of the Columbia River. Finest of
   the low evergreen shrubs we have" (Scott).

   M. japonica. "Not long grown in this country, but considered the hardiest
   and most showy" (Scott).

Paeonia moutan (now *P. suffruticosa*)   Tree Peony

   "Magnificent plants, with flowers of various shades of red, lilac, light purple
   and white. . . . I have had a plant of this with from seventy to eighty
   flowers upon it at one time, presenting a splendid sight. The flowers vary
   on the same bush: some of them are very double . . . others are semi-
   double" (Breck). "Among the most showy of the low garden shrubs"
   (Scott).

   P. moutan papaveracea (now *P. suffruticosa* var. *papaveracea*). Poppy-flowered
   tree Peony. "Large single white flowers with a purple blotch at the base of
   each petal. Very desirable plant" (Breck).

   P. moutan papaveracea rosea (now *P. suffruticosa* var. *papaveracea rosea*). "Rose
   colored flowers. Not very common" (Breck).

Philadelphus

   P. coronarius. Mock orange. Common syringa. "Delightfully fragrant when
   in bloom. The Syringa is a most delicious shrub: the foliage is luxuriant,
   the blossom beautiful and abundant, white as the purest lily, and of the
   most fragrant scent. In a room, indeed, this perfume is too powerful, but
   in the air it is remarkably agreeable" (Breck).

   P. coronarius var. *grandiflorus.* "Large flowering Syringa—also called Mock
   Orange. Handsomest of the genus" (Breck). "This old vigorous and
   graceful shrub is still one of the finest grown singly and in masses. The
   variety is not large and the common sort is still unsurpassed in fragrance"
   (Scott).

   P. hirsutus. Not fragrant. "Native of North America where it was discovered
   by Mr. Nuttall" (Breck).

Physocarpus opulifolius   Ninebark (see *Spiraea opulifolia*)

Pieris

   P. floribunda (formerly *Andromeda floribunda*) (Henderson, Breck).

   P. japonica (formerly *Andromeda japonica*).

Pinus Mugo   Mountain Pine

"Indigenous on the mountains from the Pyrenees to the Austrian Alps. A great variety of forms of this species may be seen in the New York Central Park. Most pleasing for small grounds" (Scott).

Prunus

P. amygdalus (formerly *Amygdalus pumila plena*). Japanese plum. Flowering almond. "Elegant flowers resembling small roses." "Most common early spring shrub" (Scott).

P. cerasus (formerly *Cerasus communis plena*). Sour cherry. Double forms.

P. japonica (formerly *P. sinensis*). "Small shrub but recently introduced . . . already a great favorite . . . will probably prove superior to the flowering almond" (Scott).

Rhododendron (formerly *Azalea*, which see)

R. canadense. Rhodora. "False Honeysuckle—Presents a magnificent appearance . . . with flowers . . . of a fine purple" (Breck). "A pretty companion for the flowering almond and Japan Quince" (Scott).

R. catawbiense. Mountain rosebay. "Native. Parent of varieties that are the hardiest and most beautiful in the northern states" (Scott).

R. maximum. Great laurel. American rosebay. "Flowers resemble bunches of roses. . . . one of the most magnificent [shrubs] in foliage and flower the country can boast of" (Breck). "Straggling open growth . . . shady, moist, rocky ground" (Scott).

R. ponticum. "Will withstand the winter in the open ground" (Breck). "Native of Armenia. Does best in shade. Not hardy in the northern states. Combined with . . . *R. caucasicum, R. chrysanthemum* and *R. punctatus,* British and continental gardeners originated thousands of seedlings" (Scott).

Rhodora canadense (see *Rhododendron canadense*)

Rhus

R. copallina. Black sumac. Mountain sumac. Shining sumac. "The varnished polish of the leaves, and the rich purple they assume in autumn, as well as the scarlet of the leafy heads of fruits, make this species one of the most beautiful of the genus" (Breck).

R. cotinus (now *Cotinus coggygria*). Smoke tree. Venetian sumac. "The flowers . . . have the appearance of downy silk, in light, airy masses, and as the plant is nearly covered with these graceful clusters, have some resemblance to puffs of smoke emerging from the graceful leaves" (Breck).

R. glabra. Smooth sumac. Scarlet sumac. "The flowers are disposed of in a large green head, of yellowish-green color, and agreeable fragrance. The berries are very acid and astringent. The leaves are used in tanning" (Breck).

R. typhina. Staghorn sumac. Virginia sumac. "This is one of the safe species, and highly ornamental in the shrubbery, on account of its elegant compound leaves and bunches of rich scarlet berries" (Breck).

R. vernix. Poison sumac. "The most poisonous woody plant of New England, but the most beautiful plant of the swamps" (Breck).

Ribes Currant

"A grace in the drooping habit adapts them for the borders of groups" (Scott).

R. aureum. Buffalo currant. "Golden-flowered Currant—Native of Missouri." "Fragrant Currant—Yellow fragrant flowers . . . perfuming the whole region in its neighborhood" (Breck).

R. sanguineum. "Red-flowering Currant—Rich, deep red flowers in May" (Breck).

R. sanguineum var. *flore plena.* "Double Crimson Currant—rich and striking" (Downing, Breck).

R. speciosum. "Crimson-flowering Currant—Far superior in brilliancy. Not very common" (Breck). "Fuchsia gooseberry," "appropriate border companion" (Scott).

Robinia hispida Rose Acacia

"Profusion of elegant rose-colored flowers—a very desirable species" (Breck).

Roses (see "Roses" section of Appendix)

Rubus odoratus Purple Flowering Raspberry

"This is the only ornamental variety . . . giving a charm to many solitary spots by its large, rose-like flowers" (Breck).

Sambucus canadenis American Elderberry, or Sweet Elderberry

"Its white flowers are highly esteemed, as having important medicinal qualities. An infusion of the bruised leaves is used by gardeners to expel insects from vines" (Breck). "Of considerable beauty" (Scott).

Sorbaria sorbifolia (see *Spiraea sorbifolia*) False Spiraea

Spiraea

S. argentea. "Silver-striped-leaved Spirea. Delicate species has variegated leaves" (Breck).

S. bella. "Pretty Spiraea—This is a dwarf species with pink flowers. A neat little shrub, worthy of a place in every collection" (Breck).

S. douglasii. "Mr. Douglas' Spiraea—This shrub is noticed by Mr. Downing, as a new species from California, having some resemblance to *S. tomentosa,* flowering in the same manner. . . . We were unfavorably impressed with its appearance" (Breck).

S. hypericifolia. Saint-Peter's-wreath. "Long garlands of white flowers giving the bush the appearance of being covered with a light fall of snow" (Breck).

S. laevigata. "Smooth-leaved Spiraea—White flowers and smooth leaves, somewhat fragrant" (Breck).

S. opulifolia (now *Physocarpus opulifolius*). Ninebark. "An abundance of white with rose-tinged flowers" (Breck).

S. prunifolia plena. Bridal wreath. "This charming shrub was introduced into Europe by Dr. Siebold, to whom our collections are indebted for so many novelties only to be procured with the utmost difficulty. It serves the attention of all amateurs, as well for its hardiness as its elegant habit and beautiful flowers. The Dutch traveler found it cultivated in the Japanese gardens, and supposes its native country to be Korea, or the North of China" (Breck). "One of the most common and most beautiful shrubs" (Scott).

S. reevesiana (now *S. cantoniensis*). Mr. Reeves's spiraea. "Snowy white clusters on arching stems. Elegant and desirable species" (Breck).

S. salicifolia. Queen of the meadows. "Neat white, sometimes rose-tinted flowers" (Breck).

S. sorbifolia (now *Sorbaria sorbifolia*). Pinnate-leaved spiraea. "Native of Siberia. The flowers are yellowish-white, produced in large dense panicles" (Breck).

S. tomentosa. Steeplebush. Hardhack. Silverbush. "Long tapering spire of purple flowers. This plant has very valuable astringent qualities, and is employed as a tonic in dysentery and other disorders of the system" (Breck).

S. trilobata. Three-lobed-leaved spiraea. "White flowers, leaves notched" (Breck). "Varieties recently brought into notice are so numerous we shall not attempt to name them. . . . This native shrub . . . has but recently attracted great attention as a garden shrub" (Scott).

Symphoricarpos Snowberry

"Used for winter bouquets" (Scott).

S. orbiculatus. Coralberry. Indian currant. "Inconspicuous pink flowers, brownish-purple berries" (Breck). "Red-berried Wax Berry—makes a perfect miniature tree when growing alone. 3 feet high" (Scott).

S. racemosus (now *S. albus*). Common snowberry. "Inconspicuous pink flowers but fine white berries after leaves have fallen" (Breck).

Syringa Lilac

"Large number of varieties in the nursery catalogs. For six lilacs only in addition to the common purple: *S. alba*, *S. emodi* (Himalayan variety good specimens in New York Central Park), *S. superba*, *S. josikea* (flowers

deep purple) (Hungarian), *S. Rothomagensis* (*S. chinensis*), *S. Persica alba*"
(Scott).

S. persica. Persian lilac. "Far more delicate and pretty than the common
lilacs, both in leaf and blossom" (Breck).

S. vulgaris. Common lilac. "Introduced during the reign of Henry VIII, for
in the inventory, taken by the order of Cromwell, of the articles in the
gardens of the palace of Nonsuch, are mentioned six lilacs, trees which
bear no fruit, but only a pleasant smell" (Loudon).

(Buist and Downing have all the species.)

Taxus Yew

T. baccata. English yew. "One of the prettiest of dwarf trees for small grounds"
(Scott). " *T. Baccata* and *T. Hibernia,* the English and Irish Yews, with the
Thuyas, and some other evergreens, are in the process of acclimation, and
I hope I shall be enabled hereafter to report them hardy" (Breck).

T. baccata var. *aurea.* Scott recommends.

T. canadensis. Ground hemlock. "Of no value" (Scott).

Thuja Arborvitae

T. occidentalis. American arborvitae. White cedar (Scott).

T. occidentalis 'Aurea.' "Golden seedling brought to notice by Sargent"
(Scott).

Viburnum Arrowwood.

V. acerifolium. Dockmackie. Possum haw. "All very beautiful" (Breck).

V. alnifolium. Wayfaring tree. Hobblebush. "Worthy of a place in every collection" (Breck).

V. dentatum. Southern arrowwood (Breck).

V. lentago. Sheepberry. Nannyberry. Sweet viburnum. "Clothed . . . with a
profusion of delicate showy flowers. Fragrant flowers followed by rich
blue berries hanging down among the curled leaves, which are beginning
to assume the beautiful hues of autumn. A tree of this kind makes a fine
appearance at the angle of a walk, or in the corner of a garden, as its delicacy
invites a near approach, and rewards examination" (Emerson, Downing).

V. macrocephalum. "Great clustered Snowball—This is a new and splendid
species, that has not been much, if any, cultivated in this country. M. Van
Houtte describes it as found growing in the gardens about Chusan,
China" (Downing).

V. nudum. Nannyberry (Breck).

V. opulus. European cranberry bush. Guelder rose. Snowball garden rose.
"Large white bunches of flowers like those of the Hydrangea" (Breck).

"So showy when in bloom that few even of townspeople are unfamiliar with it" (Scott).

V. opulus var. *sterilis*. Guelder rose. Snowball. "Common ornament" (Breck).

V. trilobum. Cranberry tree. Highbush cranberry. "White flowers followed by red fruit of a pleasant acid taste, resembling cranberries, for which it is sometimes substituted" (Breck).

Weigela japonica

"This noble shrub introduced from Japan as late as 1843 has already found a place in most home grounds from Maine to California" (Scott).

W. rosea (now *W. florida*). "The Rose-colored Weigela—New shrub introduced from China and first noticed by Downing." "I immediately marked it as one of the finest plants of Northern China and determined to send plants of it home in every ship, until I should hear of its safe arrival" (Robert Fortune).

Zanthorhiza apifolia   Shrub Yellowroot

"Interesting half-hardy, evergreen shrub" (Henderson). "Parsley-leaved yellow-root" (Buist).

Zenobia pulverulenta (formerly *Andromeda speciosa*)

"Flowers in great profusion" (Breck).

VIII   Trees

Trees were in demand in the seventeenth century for housing and heating, shipbuilding, barns, and fencing. In the eighteenth century they were sought for their ornamental value. In nineteenth-century landscaping their individual characteristics became of first importance. Superlatives were in order—the tallest, darkest, lightest, most bloom-laden, most oddly shaped, brightest-colored, most weeping, most contorted. Round-topped trees were chosen to stand by steep gables, trees with pointed tops to contrast with flat roof lines. Single trees were to command otherwise undisturbed sweeps of lawn. Shade trees for streets and drives were obligatory.

The enormous numbers of American native oaks, maples, and evergreens allow our listing here only those most highly recommended by the leading authorities—Joseph Breck, Frank J. Scott, George B. Emerson, and A. J. Downing. We have used their identifications in listing the trees, giving the modern botanical names in parentheses after each entry. The fact that "Spruce Fir" stood for what is now recognized as a wide variety of evergreens makes for some surprising reidentifications. For convenience in confusion we have

Autumn in New England: Cider Making.

placed the comments of these authorities under the early names by which the trees were known to them; the modern listings will refer the reader to these earlier names. We can only hope we have given enough cross-references.

Abies Fir (then "Spruce Fir")

A. alba (now *Picea glauca*). Silver fir of Virgil. European silver fir. "Similar and superior to the Balsam fir." "Single or White Spruce. Magnificent decorations for the pleasure grounds" (Breck).

A. balsamea. Balm of Gilead. Balsam fir. (See comments under *Picea balsamea*.)

A. canadensis (now *Tsuga canadensis*). "Hemlock. Has not been introduced into our pleasure grounds to any great extent" (Breck). "The most beautiful and available of all evergreens for the embellishment of small places" (Scott). "Most beautiful" (Emerson).

A. communis (now *Picea abies*). "Norway Spruce. Finer than either Black or White Spruce" (Breck). "It is so useful and valuable a tree that it is destined to become much more popular. . . . There is no planter of new places or improver of old ones who will not find it necessary to call it in for his assistance . . . strikingly picturesque . . . favourite with the artists" (Downing).

A. douglasii (now *Pseudotsuga menziesii*). "Douglas. Spruce Fir. Remarkable and desirable" (Emerson). "One of the great trees of California and Oregon" (Scott).

A. fraserii. Southern balsam. Double balsam fir. "From the great richness and luxuriance of the foliage, the double Balsam Fir is a very beautiful tree" (Emerson).

A. nigra (now *Picea mariana*). "Black Spruce. Mathematical exactness of its shape beautiful" (Breck).

A. pulcherrima (now *A. alba*). "Of Virgil. European Silver Fir. Similar and superior to the Balsam Fir" (Breck).

Acer Maple

A. platanoides. Norway maple. "More vigorous growth than the sugar and a similar formality of contour" (Scott).

A. pseudoplatanus. Sycamore maple (Scott).

A. rubrum. Scarlet maple (Scott).

A. saccharinum. White or silver-leaved maple. "Has become perhaps too great a favorite for street planting" (Scott).

A. saccharum. Sugar or rock maple. "The most valuable ornamental tree of all the maples" (Scott).

Aesculus Horse Chestnut, Buckeye

A. flava (now *A. octandra*). Sweet buckeye. "Native of western and southwestern states. Yellow flowers" (Scott).

A. glabra. Ohio buckeye. "Rough bark" (Scott).

A. hippocastanum. Horse chestnut. "Known in Europe for three centuries . . . should never be crowded . . . most perfect of single lawn trees" (Scott). "Well known ornamental tree of rapid growth" (Breck).

A. hippocastanum 'Rubricunda.' "Most showy of trees" (Scott).

A. hippocastanum var. *flore pleno*. "Superb variety with double flowers. Exquisite lawn trees . . . one of the thriftiest of the species" (Scott).

Ailanthus altissima Tree of Heaven

"Should be under ban for street planting—once popular, now in bad odor" (Scott).

Amelanchier vulgaris (now *Amelanchier ovalis*) Serviceberry, Shadbush (Scott)

Amygdalus persica plena (now *Prunus persica*)

"Double flowering Peach—Flowers large and full like small roses. . . . Unless the trees are headed down, or pruned well, they become straggling and unsightly" (Breck).

Andromeda arborea (now *Oxydendrum arboreum*) Sorrel Tree, Sourwood

"One of the prettiest additions to our stock of small ornamental trees" (Scott).

Anona triloba (now *Asimina triloba*) Custard Apple, Papaw

"Protected situation in Central Park" (Scott).

Aborvitae (see *Thuja*)

Asimina triloba (see comments under *Anona triloba*)

Betula Birch
  B. *laciniata* var. *pendula*. Cut-leaved weeping birch. "Acknowledged queen of all" (Scott).
  B. *populifolia*. Grey birch. "Nothing can be prettier seen from the windows of the drawingroom than a large group of trees whose depth and distance is made up of the heavy and deep masses of ash, oak and maple—the portion nearest to the eye on the lawn terminated by a few birches" (Downing).

Carpinus americana (now *C. caroliniana*) Blue Beech, American Hornbeam
  "Pretty isolated tree of airy outline" (Scott).
Carya alba (now *C. ovata*) Mockernut Hickory, Shagbark
  "Excels in beauty of leaf. . . . Hickories are always favorites with children. Their elastic limbs never snap treacherously" (Scott).
Castanea americana (now *C. dentata*) American Chestnut
  "Once one of the glories of rock hillsides and pastures of New England . . . well known throughout the northern states" (Scott).
Catalpa syringifolia (now *C. bignonioides*) Indian Bean, Catawba
  "Native of our southern states. Beautiful when seen singly." "Not very regular in its growth, but, when planted among other trees, or shrubs, it makes a fine appearance" (Breck).
Cedrus libani Cedar of Lebanon
  "Probably becomes a grand tree in the upper table-lands and mountains of the southern states . . . most beautiful specimens in the world planted in England in the last 200 years" (Scott). "Most celebrated tree of the genus . . . singularly bold and picturesque . . . one or two specimens will give force and character to the dullest front of round-headed trees" (Loudon). "When old and finely developed on every side, not equalled, in ornamental point of view, by any sylvan tree of temperate regions. . . . Grand and magnificent. . . . Planted in the midst of a broad lawn, it will eventually form a sublime object, far more impressive and magnificent than most of the country houses which belong to the private life of a republic" (Downing).
Cerasus (now all under *Prunus*) Cherry
  C. *pensylvanica*. Northern wild red cherry. "Delicate foliage . . . handsome white flowers . . . fruit is deep-red, and not very abundant" (Breck).
  C. *semperflorens*. "Ever flowering weeping cherry. One of the prettiest of small weeping trees" (Scott).
  C. *serotina*. Black cherry. "Well-known . . . handsome in flower and fruit . . . peculiarly subject to the ravages of the caterpillar" (Breck).
  C. *virginiana*. Chokecherry. "Grows wild all over our country. When grown

in rich soil its shining foliage contrasts well with birches, the catalpa, or the koelreuteria."

Cercis canadensis   Judas Tree, Redbud

"Indigenous to the southern United States; often seen in large collections of plants in gardens in New England" (Breck, Scott).

Chamaecyparis thyoides   White Cedar (Scott)

Chionanthus virginicus   Fringe Tree

"Native of North America. Flowers white in numerous long bunches" (Breck). "Most elegant little trees that can grace a lawn" (Scott).

Cornus   Dogwood

C. florida. Flowering dogwood. "One of the most desirable of all the genus . . . a conspicuous object, in some of our woods" (Breck).

C. mas. Cornelian cherry. Male dogwood (Scott).

Crataegus   Thorn

"It is found that a greater variety of beautiful small trees and ornamental shrubs can be formed of the several species of Thorn, than of any kind of tree whatever. Thus they give persons whose grounds are not extensive, the means of ornamenting their grounds with great facility" (Emerson).

C. coccinnea. Scarlet-fruited thorn (Breck).

C. crus-galli. Cockspur thorn (Breck, Scott).

C. oxycantha. Common hawthorn of England (Breck).

C. punctata. Dotted-fruited thorn (Breck).

C. tomentosa. Pear-leaved thorn (Breck).

Cupressus   Cypress

"By the ancients the cypress was considered an emblem of immortality; with the moderns, it is emblematical of sadness and mourning" (Emerson).

C. sempervirens. "Common Cypress of Europe—This is a tall, graceful, plumed-shaped tree, the common and suitable ornament for burying places on the Levant" (Emerson).

C. thyoides (now *Chamacyparis thyoides*). White cedar. Swamp cedar. "Graceful and beautiful . . . entirely free from the stiffness of the Pines, and to the spiry top of the Poplar, and the grace of the Cypress, it unites the airy lightness of the Hemlock" (Emerson). "We are not aware that this beautiful native tree has been cultivated for ornamental purposes; we see no reason why not" (Breck, Scott).

Cytisus laburnum (now *Laburnum anagyroides*)   Golden Chain (Scott).

Dyospyros virginiana   Persimmon

"Greatest beauty of its foliage develops farther south" (Scott).

Fagus Beech

 F. americana (now *F. grandifolia*). "American White Beech—Most common native species" (Scott).

 F. sylvatica. European beech (Scott).

 F. sylvatica var. *pendula.* "Weeping Beech—The most curious beech of our zone. The original tree stands in the park of Baron de Man in Beersel, Belgium" (Scott). "Beeches are suitable for ornamental pleasure grounds, but too large for shrubbery. No collection . . . should be deficient of the Purple, or Copper Beech. . . . Weeping Beech is another desirable variety" (Breck).

 F. sylvatica var. *purpurea.* "Sport from common white beech found in a German forest" (Scott).

Fraxinus Ash

 F. americana. American ash. "Well-known valuable timber tree, and suitable for avenues, but not for the shrubbery, unless on a large scale" (Breck).

 F. americana var. *pendula* (now *F. rotundifolia* var. *pendula*). "Weeping Ash— Branches will droop to the ground and form a handsome weeping head" (Breck).

 F. excelsior. European ash (Breck).

Ginkgo Maidenhair Tree

 Salisburia adiantifolia (now *G. biloba*). "Recommended on high English authority." Introduced about 1730 (Scott).

Halesia Silver Bell, Snowdrop Tree

 H. diptera. "Less hardy. South Carolina. Tennessee" (Scott).

 H. tetraptera (now *H. carolina*). "Fine old specimen in New York Central Park" (Scott).

Ilex opaca American Holly

 "Has considerable beauty . . . particularly valuable for retaining its bright green leaves through the year, and for its berries" (Emerson).

Juniperus Virginia Cedar, Red Cedar

 J. cupressa thyoides (now *Chamaecyparis thyoides*). White cedar (Scott).

 J. virginiana. "Red Cedar—Savin. Although so common and monotonous in its appearance on the rocky shores of Massachusetts, it may be introduced with good effect among other evergreen trees" (Breck). "Just falls short of being one of the most beautiful of evergreens" (Scott).

Koelreuteria Golden Rain Tree
"Hardy—From northern China. Introduced into England in 1763. Long cultivated in the United States but little known" (Scott).

Larix Larch, Tamarack
"Deciduous trees of mountainous regions of both continents. For ornamental purposes the Larch is important, on account of its rapid growth, beautiful symmetrical shape, and thick foliage, which is of an agreeable light bluish-green" (Breck).

L. americana (now *L. laricina*). Tamarack. Hackmatack. "Inferior to the European" (Breck, Scott).

L. communis (now *L. decidua*). European. "Found growing in company with evergreens" (Scott, Breck).

L. europaea (now *L. decidua*). "Scotch Larch—Much over-valued for ornamental purposes" (Scott).

L. europaea pendula (now *L. decidua* 'Pendula'). "Curious and valuable, picturesque small tree" (Scott, Breck).

Maclura pomifera Osage Orange
Native of Missouri and Arkansas. "Beauty of the tree itself is sufficient" (Scott). As a shrub (see "Shrubs" section), much used for hedges.

Magnolia
M. acuminata. Cucumber tree. "In . . . Boston there are handsome specimens of this magnificent tree, but not of a large size" (Breck).

M. auriculata. Ear-leaved magnolia. "Splendid tree . . . beautiful for shape, foliage and flowers" (Breck).

M. conspicua. White magnolia. "The most charming of all Magnolia. The flowers . . . a pure, creamy white, are produced in such abundance that the tree . . . may be seen at a great distance. . . . 'The Lily Tree' . . . among the Chinese poets considered the emblem of candor and beauty" (Downing).

M. soulangiana. "Outside petals finely tinged with purple" (Downing). Downing suggests planting American arborvitae and Hemlock as a dark green background "on which the beautiful masses of Magnolia flowers would appear to great advantage . . . since they bloom before the leaves appear."

Morus Mulberry
M. alba. White mulberry. "Specimens . . . in New York Central Park . . . among most beautiful to be seen" (Scott).

M. rubra. Red mulberry. "American Red Mulberry—Largest and finest ornamental tree of the genus . . . Most domestic in expression" (Scott).

Oxydendrum arboreum (for comments, see *Andromeda*) Sorrel Tree, Sourwood

Paulownia imperialis (now *P. tomentosa*)
"Japanese tree introduced into France in 1837. Fine specimen in New York Central Park" (Scott). "Magnificent tree of recent introduction. Showy foliage . . . leaves . . . resembling those of a large sun-flower" (Breck).
Persimmon (see *Dyospyros virginiana*)
Picea Spruce
"Suitable for ornamenting the shrubbery or lawn, when planted in groups, but not proper for single specimens" (Breck).
P. abies. Norway Spruce (for comments, see *Abies communis*).
P. balsamea (now *Abies balsamea*). Balsam fir. Balm of Gilead. "Possesses in a high degree the most important qualities of the evergreens, as an ornamental tree—regular pyramidal shape, and rich, deep-green foliage" (Emerson). "Does not grow old gracefully" (Scott).
Pinus Pine
P. austriaca (now *P. nigra*). "Austrian Pine—Well deserving attention" (Downing).
P. pungens. Prickly pine. "Table Mountain Pine. Described by Michaux" (Scott).
P. resinosa. "Red or Norway Pine—More ample in its dimensions. We should recommend this species only where there are extensive grounds to decorate" (Breck).
P. strobus. "White Pine—Familiar to all . . . growing to stately size" (Breck). "We place the White Pine among the first in the regards of the ornamental planter" (Downing).
P. sylvestris. "Scotch Pine—Found in the British Islands. More claim to beauty than the Pitch Pine (*P. rigida*). A few trees . . . admissible . . . in large plantations" (Breck).
P. taeda. "Loblolly Pine—No value north of Virginia . . . peculiar to the sand barrens of the southern states" (Scott).
Prunus persica (for comments, see *Amygdalus persica plena*)
Pseudotsuga menziesii (for comments, see *Abies douglasii*)
Pyrus (now Pear, Apple)
P. americana (now *Sorbus americana*). "American Mountain Ash—More dwarf and bushy than the European" (Breck).
P. aucuparia (now *Sorbus aucuparia*). Rowan. Quickbeam. "European Mountain Ash—More graceful in its habits than American species. Berries . . . constitute the great ornament of the tree. Commonly known in England by the name of Rowan or Roan-tree" (Breck).

P. coronaria (now *Malus coronaria*). "Sweet-scented Crab—Native of North America. When it flowers, a delightful fragrance is emitted which in the evening perfumes the whole of that part of the garden" (Breck).

P. prunifolia (now *Malus prunifolia*). "Siberian Crab—Ornamental in flower and fruit. Fruit sometimes used as a preserve" (Breck).

Quercus alba

"White Oak—Grandest, most common and most useful of our northern oaks" (Scott).

Rhamnus cathartica   Common Buckthorn

"Fruit . . . formerly employed, in medicine, as a purgative, it is too violent and drastic to be safely used, and is now chiefly confined to veterinarian practice, to which it is well adapted. The saffron-colored juice of the unripe berries, called French berries by dyers, is used as paint and a dye. Sap-green is made of the inspissated juice of the ripe berries, with alum and gum Arabic. If gathered very late, they yield a purple, instead of a green, color. The bark furnishes a beautiful yellow dye, or, dried, it colors brown" (Emerson).

Robinia   Locust

R. crispa. "Crisp-leaved Robinia—This with following list imported last year: *R. inermis; R. tortuoso,* the branches all growing in a circular, zigzag style; *R. macrophylla, R. sophorafolia, R. volubilis, R. elegans* and *R. grandiflora.* Singularly curious and elegant leaves" (Breck).

R. pseudoacacia. "Common Locust—Too well known to make it necessary to give a description. . . . Blossoms are butterfly or pea-shaped, white, with yellow in the middle" (Breck).

R. viscosa. "Clammy-barked locust—Small tree, large pale-pink flowers . . . Looks well with other trees and shrubs" (Breck).

Sophora japonica   Pagoda Tree

"Curious so few yet to be seen in this country" (Scott).

Sorbus (for comments, see *Pyrus*)

S. americana. American mountain ash.

S. aucuparia. European mountain ash.

Thuja   Arborvitae, cedar

T. occidentalis. American white cedar. "American Arbor Vitae—Remarkable for its graceful pyramidal, spire-like shape, thickly clothed with branches from the ground to the apex. Forms a very ornamental hedge, and is coming very much into use for protecting gardens from cold" (Breck).

Ulmus Elm
> *U. americana.* "American Weeping or White Elm—Certainly the queen as the oak is the king among deciduous trees" (Scott).
> *U. campestris* (now *U. carpinifolia*). English elm. Smooth-leaved elm. "Finest trees of this species we have seen in this country are on the Boston Common" (Scott).
> *U. glabra.* Scotch elm. Wych elm. Camperdown elm. "Comparatively new variety. Deep shady bower of novel beauty. A valuable addition to our stock of gardenesque trees" (Scott).

IX *Vines*

As vines were considered an adjunct to every garden—in that they would climb the sides of houses, cover an arbor, form a screen between house and service quarters, shade a summerhouse, cover a porch, and climb trees—it is no wonder the nineteenth-century gardeners had their special favorites. Apart from grapes, vines were admired for their fancy leaves, colored fruits, fragrant blooms, and odd seedpods. The vines here given are all woody and deciduous except for the climbing euonymus, which may be considered evergreen.

Downing saw vines as the "draperies" of a house. In addition to using them on walls and trellises, he planted vines to run up dead cedar stumps or over trunks.

Where there is confusion, we have left the nineteenth-century names as listed by the users and modernized them where necessary, leaving their comments under the names they used.

Adlumia fungosa Climbing Fumitory, Mountain Fringe, Allegheny Vine
> "Elegant. Indigenous" (Breck).

Akebia quinata Five-Leaved Akebia, Chocolate Vine
> "Recently introduced . . . already quite a favorite" (Scott). "Purple, fragrant" (Henderson).

Ampelopsis quinquefolia (now *Parthenocissus quinquefolia*) American Woodbine
> "The most ornamental plant of its genus" (Emerson, Buist). "Examples of the surprising luxuriance of this plant may be seen on a number of dwelling houses in Beacon Street, Boston, and on many other buildings in that city" (Breck). Virginia creeper—"leaves grandly colored in fall. Used on railroad embankments to prevent sliding" (Henderson).

Aristolochia sipho (now *Aristolochia durior*) Dutchman's Pipe, Birthwort
> "A singular climbing plant, with handsome, broad foliage, with brownish purple and very curious somewhat pipe-shaped flowers" (Breck). "Twiner

and climber . . . wild in the middle states. Finest exhibition of massy foliage for covering constructions" (Scott). "Dutchman's Pipe—greenish brown, curious" (Henderson). "Excellent arbour twiner" (Buist).

Bignonia Trumpet Flower, Cross Vine

    B. atrosanguinea (now *B. capreolata* var. *atro sanguinea*). (Scott, Buist).

    B. grandiflora (now *Campsis grandiflora*). "Succeeds in a more southern climate . . . in north with some protection" (Breck, Buist). "Chinese variety [formerly *Campsis chinensis*] not quite so hardy" (Scott).

    B. radicans (now *Campsis radicans*). Scarlet trumpet flowers. "Magnificent climbing plant . . . of great beauty" (Breck). "Common Trumpet Creeper—Superb vine to grow on old evergreen trees" (Scott).

Calampelis scaber (now *Eccremocarpus scaber*)

    "Rough-podded calampelis—A beautiful climber. Bright orange color; . . . flowers profusely the latter part of summer" (Breck).

Campsis Trumpet Creeper, Crown Plant

    C. grandiflora (for comments, see *Bignonia*).

    C. radicans (for comments, see *Bignonia*).

    C. tecoma (for comments, see *Tecoma*).

Caprifolium (now *Lonicera*) Honeysuckle

    C. belgicum (now *Lonicera Periclymenum*). "Dutch Sweet-scented Honeysuckle—Profusion of bloom . . . which emits a delightful odor to all the neighborhood" (Breck).

    C. flavum (now *Lonicera flava*). "Yellow Trumpet Monthly Honeysuckle—Native of South, but long cultivated in Europe, and from there introduced here" (Breck).

    C. flexuosum (now *Lonicera japonica*). "Chinese Honeysuckle—Evergreen leaves . . . blooms with exhaustless profusion" (Breck).

    C. floribunda. "One of the beautiful varieties of the Scarlet Trumpet Honeysuckle . . . imported by us a few years since, that have given great satisfaction" (Breck).

Cardiospermum halicacabum Balloon Vine, Heartseed, Love-in-a-Puff (Breck)

Celastrus scandens Wax-work, Climbing Staff, Bittersweet

    "This native climber should be introduced into every garden" (Breck). "Bittersweet—of little value for culture" (Scott).

Clematis Virgin's Bower

    C. azurea. "All shades from white to deepest purple and blue, double and single" (Henderson, Buist).

    C. azurea grandiflora (Now *C. florida*). "Great-flowering Blue Virgin's Bower—

Larger flowers than the variety Sieboldii" (Breck, Buist). "Chinese . . . not long introduced . . . Flowers larger than the Native or European sorts" (Scott, Downing).

C. *flammula*. "Luxuriant climber, producing clusters of small white flowers" (Breck, Buist). "European sweet-scented" (Scott, Downing). "Virgin's Bower—white, fragrant" (Henderson).

C. *florida*. "Large, white flowers . . . luxuriant" (Breck). "Introduced before 1798" (Buist). "Japanese species" (Scott).

C. *virginiana*. Virginia bower. "Profuse and conspicuous white flowers" (Scott, Buist).

C. *vitalba*. Traveler's joy. "Old Man's Beard—a useful vine to cover unsightly roofs" (Scott).

C. *viticella*. "More showy, larger flowers" (Scott).

Eccremocarpus scaber (for comments, see *Calampelis scaber*)

Hedera

H. *canariensis*. Irish ivy. "*H. canariensis variegata* has silver striped and gold striped leaves" (Scott).

H. *helix*. English ivy. "This in its many varieties is scarcely hardy at New York" (Henderson). "Common Ivy—Much used in England for covering naked buildings or trees, or for training into fanciful shapes, or trained up a stake so as to form a standard. In this country it is not very common. . . . Some specimens in . . . Boston . . . flourish finely upon the rough granite or brick walls of buildings" (Breck).

Ipomoea Morning Glory

I. *purpurea*. Common morning glory.

I. *quamoclit*. Cypress vine. Cardinal climber. Universally recommended. "No annual climbing plant exceeds this" (Breck).

Lonicera (see also *Caprifolium*) Honeysuckle

L. *halliana* (variety of L. *japonica*). Hall's Honeysuckle, white and buff" (Henderson).

L. *japonica*. "Japan Honeysuckle, pink and white" (Henderson, Downing). "Tender north of Southern Virginia" (Buist).

L. *japonica* var. *aurea reticulata*. "Golden-leaved honeysuckle" (Henderson).

L. *periclymenum*. "English Woodbine—crimson, shaded white" (Henderson). "Most showy in its flowers" (Scott).

L. *periclymenum* var. *belgica*. "Dutch Honeysuckle, monthly rose and white" (Henderson, Downing).

L. periclymenum 'Flava.' "Bright yellow, native of our southern states" (Scott, Downing).

L. sempervirens. Trumpet honeysuckle. "Scarlet flowers, sub-evergreen in the South" (Scott). "Our coral honeysuckle" (Buist).

Lycium barbarum   Boxthorn, Matrimony Vine

"Willow-leaved Lycium—"Ornamental climbing shrub, valuable for covering arbors, naked walls, etc. . . . delicate foliage . . . plant covered with small handsome violet flowers. In the shrubbery, it may be permitted to ramble at its pleasure, or trained to suit the fancy" (Breck).

Maurandia lophospermum (now *Asarina lophospermum*)

"Handsome plant" (Breck).

Parthenocissus quinquefolia (for comments, see *Ampelopsis quinquefolia*) American Woodbine

Passiflora incarnata   Passion Vine, Maypop

"Most beautiful" (Buist).

Periploca graeca

"Virginia Silk Vine—Odor said to be unwholesome . . . should not be planted on porches or near windows" (Scott).

Phaseolus coccineus   Scarlet Runner Bean (Downing, Buist)

P. *multiflora* (Downing, Buist).

P. *multiflorus.* "Scarlet-flowering bean—spikes of showy scarlet flowers, and a variety with white flowers . . . cultivated to cover arbors, walls, or to form screens" (Breck).

Quamoclit (now *Ipomoea quamoclit*)   Cypress Vine

"No annual climbing plant exceeds this" (Breck).

Rhus toxicodendron

"Poison Ivy—Handsome . . . would be desirable for covering walls, trees, etc., were it not for its poisonous qualities" (Breck).

Tecoma (now *Campsis*)   Trumpet Bush, Merrybells

T. *grandiflora* (now *Campsis grandiflora*). "Native of China and Japan. Flowers cup-shaped . . . orange-scarlet, marked with streaks . . . brilliant appearance for a long time. It is well worthy the attention of those who are looking for climbers of a permanent kind to cover unsightly walls or close fences, or to render garden buildings of any kind more ornamental by a rich canopy of foliage and bloom" (Downing).

T. radicans (now *Campsis radicans*). Trumpet creeper. "Orange" (Henderson).
Thunbergia   Clock Vine
  T. alata. Black-eyed Susan vine. "Winged petioled thunbergia—Numerous buff-colored flowers, with dark throat" (Breck).
  T. alata 'Alba.' "White flowered" (Breck).
  T. alata 'Aurantiaca.' "Orange flowered." "There are other improved varieties, all fine. The genus was dedicated by the younger Linnaeus to his friend and successor, Thunberg, an indefatigable botanical traveller" (Breck).

Wisteria
  W. frutescens. American wisteria. (Scott, Buist). "Purple and white" (Henderson).
  W. frutescens var. *magnifica* (now *W. macrostachya*). "Magnificent Wisteria—Dark purple, Large" (Henderson).
  W. sinensis. Chinese wisteria. "Purple" (Henderson). "Magnificent . . . thousands of rich clusters of pendulous racemes of delicate pale purple blossoms" (Breck, Downing, Buist).
  W. sinensis 'Alba.' "White Wisteria" (Henderson). "Brought to England from China by Mr. Fortune" (Breck). "Recently imported" (Scott).

Sources for Plant List

Bailey, L. H. *Hortus Third*. New York: Macmillan, 1976.

Breck, Joseph (seedsman and florist, former editor of *The New England Farmer* and *The Horticultural Register*). *The Flower Garden; or, Breck's Book of Flowers*. New York: A. O. Moore, Agricultural Book Publishers, 1859.

———. *The New Book of Flowers*. New York: Orange Judd, 1866.

Bridgeman, Thomas (gardener, seedsman, and florist). *The Young Gardener's Assistant*. "For sale by the author, Broadway, Corner of Eighteenth Street," 1847.

Buist, Robert (nurseryman and florist). *The American Flower Garden Directory*. Philadelphia: Carey and Hart, 1839.

Downing, A. J. *Cottage Residences*. 3rd ed. New York: Wiley and Putnam, 1847.

———. *Landscape Gardening, . . . adapted to North America; with a view to the improvement of country residences*. New York: G. P. Putnam, 1849.

———, ed. *Gardening for Ladies, and Ladies' Companion to the Flower Garden*, by Mrs. Loudon. New York: John Wiley, 1853.

Henderson, Peter. *Handbook of Plants*. New York: Peter Henderson, 1881.

Johnson, G. W. *The Cottage Gardener's Dictionary*. 5th ed. London: W. Kent, 1860.

Parkman, Francis. *The Book of Roses*. Boston: J. E. Tilton, 1866.

Scott, F. J. *The Art of Beautifying Suburban Home Grounds*. New York: Appleton, 1870.

Bibliography

As in my other books, for those who prefer bibliographies to texts, I indicate sources with these abbreviations:

A.	Author's Library
B. A.	Boston Athenaeum
C. H. S.	Cambridge Historical Society
E. I.	Essex Institute
H. B. L.	Hunt Botanical Library
H. H.	Harvard Herbarium
I. P. L.	Ipswich Public Library
M. H. S.	Massachusetts Horticultural Society
P. A.	Portsmouth Athenaeum
P. H. S.	Pennsylvania Horticultural Society

Friends are listed by name

Bailey, Liberty Hyde. *The Standard Cyclopedia of Horticulture*. New York: Macmillan, 1950.   A.

————. *Hortus Third*. New York: Macmillan, 1976.   I. P. L.

Beecher, Catherine, and Stowe, Harriet Beecher. *The American Woman's Home*. 1855; reprinted by the Stowe Foundation from 1869 edition.   A.

Benjamin, Asher. *The American Builder's Companion*. 3rd ed. 1814; Dover Reprint, 1965.   A.

Bergen, Joseph Y., and Davis, Bradley M. *Principles of Botany*. Boston, New York: Ginn, 1906.   A.

Bicknell and Co. *Victorian Buildings*. 1875; Dover Reprint, 1979.   A.

Bicknell, A. J. *Victorian Architecture: 1873 Pattern Book*. Introduction by John Maas. Reprinted by the American Life Foundation and Study Institute, New York, 1978.   A.

Bigelow, Jacob. *Florula Bostoniensis; or, A Collection of the Plants of Boston and Its Vicinity*. 2nd ed. Boston: Cummings Hilliard, 1824.   A.

————. *Florula Bostoniensis.* 3rd ed., enlarged. Boston: Little, Brown, 1840.
Pattie Hall
————. *A History of the Cemetery of Mount Auburn.* Boston: J. Munroe, 1860.
H. H. S.
British Husbandry: Farming for Ladies. London: John Murray, 1844.   A.
Cleaveland, H. W. S. *Landscape Architecture, 1873.* Reprint. Pittsburgh: University of Pittsburgh Press, 1965.   Thomas Kane
Cleaveland, Henry W.; Backus, William; and Backus, Samuel D. *The Requirements of American Village Homes.* Reprint. New York: Appleton, 1856. A.
Cobbett, William. *The American Gardener.* London: Clement, 1821.   A.
Colonial Society of Massachusetts. *Medicine in Colonial Massachusetts, 1620–1820.* Boston: Colonial Society of Massachusetts, University Press of Virginia, 1980.   A.
Curtis, William. *The Botanical Magazine; or, Flower Garden Displayed.* London, 1787–1800.   B. A.
Cutright, Paul Russell. *Lewis and Clark, Pioneering Naturalists.* Urbana: University of Illinois Press, 1969.
Darby, John. *Botany of the Southern States in Two Parts. I. Structural and Physiological Botany. II. Description of Southern Plants Arranged in Natural System.* New York: Barnes and Burr, 1860.   A.
Darlington, William. *Memorials of John Bartram and Humphry Marshall.* Introduction by Joseph Ewan. New York: Hafner, 1967.   A.
Darwin, Erasmus. *The Botanic Garden: A Poem in Two Parts. I. The Economy of Vegetation: II. The Loves of the Plants, with Philosophical Notes.* New York: T and J Swords, Printers to the Faculty of Physic of Columbia College, 1790.   E. I.
Dictionary of American Biography.   M. H.
Dictionnarie de Traite Universel des Drogues Simples. Amsterdam, 1714.   A.
Downing, A. J. *Cottage Residences.* 3rd ed. New York: Wiley and Putnam, 1847.   A.
————. *Landscape Gardening, a treatise on the theory and practice of landscape gardening, adapted to North America; with a view to the improvement of county residences . . . with remarks on Rural Architecture.* 4th ed., revised and enlarged. New York: G. P. Putnam, 1859.   A.
————. *Rural Essays.* New York: R. Worthington, 1881.   A.
————, ed. *The Horticulturist.* Vol. 2 (1847–48).   A.
Edmonds, Walter D. *The Erie Canal.* Lunenburg, Vt.: Stinehour Press, 1960.
Walter Edmonds
Edwards, Sydenham. *The Botanical Register.* London, 1815.   M. H. S.
Eifert, Virginia S. *Tall Trees and Far Horizons.* New York: Dodd, Mead, 1965.   A.

Elder, Walter. *The Cottage Garden of America*. 2nd ed. Philadelphia: Moss, 1854.   June Hutchinson

Eliot, Charles W. *Charles Eliot*. Boston: Houghton Mifflin, 1902.   P. A.

Emerson, George B. *Report on the Trees and Shrubs Growing Naturally in the Forests of Massachusetts*. "Published Agreeably to an Order of The Legislature." Boston: Dutton and Wentworth, 1846.   A.

———. *A Report on the Trees and Shrubs Growing Naturally in the Forests of Massachusetts*. 4th ed. 7 vols. Boston: Little, Brown, 1887.   A.

Erichsen-Brown, Charlotte. *Use of Plants for the Past 500 Years*. Aurora, Ont.: Breezy Creeks Press, 1979.   A.

Estes, J. Worth. *Hall Jackson and the Purple Foxglove: Medical Practice and Research, 1760–1820*. Hanover: University Press of New England, 1979. A.

Ewan, Joseph. Introduction to *Memorials of John Bartram and Humphry Marshall*, by William Darlington. New York: Hafner, 1967.   A.

Fein, Albert. *Frederick Law Olmsted and the American Environmental Tradition*. New York: George Braziller, 1972.   M. H. S.

———. *Landscape into Cityscape: Frederick Law Olmsted's Plans for a Greater New York City*. Ithaca, N.Y.: Cornell University Press, 1980.   M. H. S.

Fessenden, Thomas Green. *The Complete Farmer and Rural Economist*. 5th ed. Boston: Otis, Broaders, 1840.   M. H. S.

———. Description of Parmentier's Garden, in *The American Farmer* (Sept. 1829), pp. 243, 260.   G.

———. *The New American Gardner, Containing Practical Directions for the Culture of Fruits and Vegetables, including Landscape and Ornamental Gardening, Grapevines, Silk, Strawberries*. Boston: Otis, Broaders, 1845.   M. H. S.

Fowler, Orson S. *The Octagon House: A Home for All; or, The Gravel Wall and Octagon Mode of Building*. 1853; Dover Reprint.   A.

Freeling, Arthur. *Flowers; Their Use and Beauty, Language and Sentiment*. London: Darton, 1857.   A.

Fuller, Margaret. *Woman in the Nineteenth Century*. London, 1845.   A.

Furnas, J. C. *Fanny Kemble*. New York: Dial Press, 1902.   I. P. L.

Gerard, John. *The Herball; or, General Historie of Plantes*. London, 1633.   A.

Gilpin, William. *Practical Hints upon Landscape Gardening*. 2nd ed. Edinburgh, 1835.   A.

Gorer, Richard. *The Growth of Flowers*. London: Faber and Faber, 1978.   A.

Gozzaldi, Isabella. *A Child in a New England Garden*. Paper read before Cambridge Plant Club. Cambridge Historical Society, 1933.   C. H. S.

Grandville. *Flowers Personified, being a translation of "Les Fleurs Animées" by Grandville*. Translated by N. Cleaveland, Esq. "Illustrated with steel engravings beautifully colored." New York: Martin, 1847.   E. I.

Graustein, Jeanette E. *Thomas Nuttall, Naturalist: Explorations in America, 1808–1841.* Cambridge: Harvard University Press, 1967.   A.

Gray, Asa. *Botany for Young People.* New York: American Book Co., 1850. A.

———. *Botany for Young People and Common Schools: How Plants Grow, with Popular Flora and 500 Wood Engravings.* New York: Furson, Phinney, Blackeman, 1865.   A.

———. *Botany for Young People and Common Schools.* 300 wood engravings. 1869.   E. I.

———. *Botany for Young People and Common Schools; How Plants Grow, with Popular Flora and 500 Wood Engravings.* New York: Ivison, Blakeman, Taylor, 1874.   A.

———. *A Manual of the Botany of the Northern United States, Arranged according to the Natural System.* Boston: Munroe, 1848.   A.

———. *Manual of Botany.* 8th ed., expanded by Merritt Lyndon Fernald. New York: American Book Co., 1950.   A.

———. *School and Field Book of Botany,* "Lessons in Botany," "Field, Forest, and Garden Botany" in one volume. New York: Ivison, Blakeman and Taylor, 1857, 1868.   A.

———. *Scientific Papers.* Selected by Charles Sprague Sargent. 2 vols. Boston: Houghton Mifflin, 1889.   E. I.

Great Trans-continental Tourist's Guide. New York: George Crofutt, 1870.   A.

Greenaway, Kate. *Language of Flowers.* Illustrated by Kate Greenaway. London: Morrison, 1977.   A.

Griffith, R. Egglesfield. *Medical Botany: More important plants used in medicine, their history, properties and mode of administration.* Philadelphia: Lea and Blanchard, 1847.   A.

Hall, Capt. Basil. *Travels in North America, 1827–1828.* 3 Vols. Edinburgh: Cadell, 1829.   B. A.

Hall, Mrs. Basil. *An Aristocratic Journey, Being the outspoken Letters of Mrs. Basil Hall written during a Fourteen Month Sojourn in America, 1827–1828.* Edited by Pope-Hennessy. New York: Putnam, 1931.   A.

Hand Book of the Sentiment and Poetry of Flowers. 2nd ed. Boston: Saxton and Kelt, 1845.   A.

Hardlin, David P. *The American Home, 1815–1915.* Boston: Little, Brown, 1979.   Julia Smith

Hare, Caspari, and Rusby. *National Standard Dispensatory.* Philadelphia: Lea, 1905.   A.

Hendrick, U. P. *A History of Horticulture in America to 1860.* New York: Oxford University Press, 1950.   A.

Henderson, Peter. *Gardening For Pleasure.* Jersey City, 1875.   A.

———. *Gardening for Pleasure.* New York: O. Judd Co., 1887.   A.

————. *Gardening for Pleasure*. New enlarged ed. New York: Judd, 1888.   A.

————. *Gardening for Pleasure*. New enlarged ed. New York: Judd, 1889.   A.

————. *Gardening for Pleasure*. 2nd ed. New York: Judd, 1901.   A.

————. *Gardening for Profit*. New York: Judd, 1866.   A.

————. *Handbook of Plants*. New York: Peter Henderson, 1881.   A.

————. *Practical Floriculture*. New York: Judd, 1869.   A.

————. *Practical Floriculture*. New enlarged ed. New York: Judd, 1887.   A.

Historical Atlas of Halton County, Ontario. Facsimile. Toronto: Walker and Miles, 1877.   A.

Holmes, Oliver Wendell. *Medical Essays*. Vol. 9. Boston: Houghton Mifflin (Riverside), 1842–82.   A.

Hutchinson, B. June. "A Taste for Horticulture," (article on Charles Mason Hovey). *Arnoldia* 40, no. 1 (Jan./Feb. 1980).   A.

Hyams, Edward. *Capability Brown and Humphry Repton*. London: Denva Sons, 1971.   A.

Jack, Annie L. *The Canadian Garden: A Pocket Help for the Amateur*. Toronto: Musson; Montreal: A. P. Chapman, 1910.   A.

Johnson, G. W. *The Cottage Gardener's Dictionary* (1846). 5th ed. London: W. Kent, 1860.   A.

Johnson, Mrs. S. O. *Every Woman Her Own Flower Gardener: A Handy Manual of Flower Gardening for Ladies*. 7th ed. New York: Ralph H. Waggoner, 1889.   A.

Kaempfer, Englebert. *The History of Japan*. 3 vols. Glasgow: James Maclehose, n.d.   Homer White

Kastner, Joseph. *A Species of Eternity*. New York: E. F. Dutton (Paperback), 1970.   Thomas Kane

Kemp, Edward. *How to Lay Out a Garden*. 3rd ed. London: Bradbury and Evans, 1864.   A.

Kern, G. M. *Practical Landscape Gardening, with Reference to the Improvement of Rural Residences, Giving the General Principles of the Art with Full Directions for Planting Shade Trees, Shrubbery, and Flowers and Laying Out Grounds*. Cincinnati: Moore, Wilstach, Keys, 1855.   A.

Kreig, Margaret B. *Green Medicine: The Search for Plants that Heal*. Chicago: Rand McNally, n.d.   A.

Kushner, Suzette. "André Parmentier in America, 1824–1830." MS given by author to M. H. S.   M. H. S.

The Language of Flowers Poetically Expressed, being a Complete Flora's Album. New York: Leavitt and Allen, 1847.   A.

La Penta, Barbara Luigia, trans. *The American City*, by Giorgio Ciucci, Francesco Dalco, Mario Manieri-Ella, Manfredo Tafuri. Cambridge: M.I.T. Press (Paperback), 1983.   Melanie Simo

Lemmon, Kenneth. *The Golden Age of Plant Hunters.* London: Phoenix House (Aldine Paperback), 1965.   A.

Lewis, Meriwether, and Clark, William. *Journals.* Edited by Bernard de Voto. Boston: Houghton Mifflin (Paperback), 1953.   A.

Lincoln, Mrs. Almira A. *Familiar Lectures on Botany.* "For the use of Seminaries and Private Students." 5th ed. New York: Huntington and Savoy, 1844.   A.

Loudon, J. C. *Encyclopaedia of Gardening.* 2 vols. New ed., improved and enlarged. London: Longman, Rees, Green, Brown, Green and Longman, 1835.   A.

———. *Encyclopaedia of Gardening.* New ed., edited by Mrs. Loudon. 2 vols. London: Longman, Green, 1871.   A.

———. *Encyclopaedia of Trees and Shrubs.* Abridgement of *Arboretium et Fruticetum Britannicum,* "Adapted for the use of Nurserymen, Gardeners, and Foresters." London: Longman, Brown, Green and Longman, 1842.   A.

———. *Encyclopaedia of Plants.* London: Longman, Rees, Green, Brown and Green, 1829.   A.

———. *The Landscape Gardening and Landscape Architecture of the Late Humphry Repton, Esq.* London: Longman, 1840.   A.

———. *The Villa Gardener.* Edited by Mrs. Loudon. 2nd. ed. London: Orr, 1850.   Joan Purington

Loudon, Mrs. *Gardening for Ladies, and Ladies' Companion to the Flower Garden.* Edited by A. J. Downing. New York: John Wiley, 1853.

———. *Gardening for Ladies, and Ladies' Companion to the Flower Garden.* Edited by A. J. Downing. New York: Wiley and Putnam, 1843.   A.

———. *The Ladies' Country Companion; or, How to Enjoy a Country Life Rationally.* 2nd ed. London: Longman, 1846.   M. H. S.

———. *My Own Garden; or, The Young Gardener's Year Book.* London: Kerby and Son, 1855.   M. H. S.

McKelvey, Susan Delano. *Botanical Exploration of the Trans-Mississippi West, 1780–1850.* Boston: Arnold Arboretum of Harvard University, n.d.
                                                                 Sedgwick Library

McKensey, Elizabeth. *Niagara Falls.* London: Cambridge University Press, 1975.   A.

M'Mahon, Bernard. *The American Gardener's Calendar.* Philadelphia: B. Graves, 1806.   A.

Malone, Dumas. *Jefferson and His Time.* Boston: Little, Brown, 1981.   A.

Marsh, George Perkins. *Man and Nature.* Edited by Davis Lowenthal. Cambridge: Harvard University Press (Belknap Press), 1964.   Melanie Simo

Martineau, Harriet. *The Martyrage of the United States.* Boston: Weeks, Jordan, 1839.   I. P. L.

———. *Society in America*. 2 vols. 4th ed. New York, London: Saunders and Otley, 1837.   I. P. L.

Mead, Peter B., Secretary, N.Y. Horticultural Society. In *The Requirements of American Village Homes, by Cleaveland, Backus & Backus*. New York: Appleton, 1856.   A.

Miller, Amy Bess. *Shaker Herbs*. New York: Potter, 1976.   A.

Miller, Phillip. *The Gardener's Dictionary*. 8th ed. London, 1768.   A.

Millspaugh, Charles F. *American Medicinal Plants*. With revised classifications. Dover Reprint, 1974.   A.

———. *Medicinal Plants Indigenous to and Naturalised in the United States*. Vol. 2. Philadelphia: John C. Yorston, 1892.   A.

Napheys, George H. *The Prevention and Cure of Disease: The Model Sick Room*. Chicago: Holland, 1871.   A.

Nuttall, Thomas. *A Journal of Travels into the Arkansa Territory during the Year 1819*. Philadelphia: T. H. Palmer, 1821.   H. B. L.

———. *North American Sylva: Forest Trees of the United States, Canada, and Nova Scotia not described in the Work of F. Andrew Michaux. 3 Vols. Being the 4th Vol. of Michaux & Nuttall's "North American Sylva."* 1857.   H. B. L.

Olmsted, Frederick Law. *Civilising American Cities*. A Selection of Writings, edited by S. B. Sutton. Cambridge: M.I.T. Press (Paperback) 1979.   A.

———. *Frederick Law Olmsted, Landscape Architect, 1822–1903*. Edited by Frederick Law Olmsted, Jr., and Theodora Kimball. New York: G. P. Putnam (Knickerbocker Press), 1922.   M. H. S.

———. *A Journey in the Back Country in the Winter of 1853–4*. 2 vols. 1860; G. P. Putnam (Knickerbocker Press), 1907.   M. H. S.

———. *A Journey in the Seaboard Slave States, 1853–54*. 1856; New York: Putnam, 1904.   M. H. S.

———. *Papers*. Vol. 1: *The Formative Years*. Edited by Charles Capen McLaughlin and Charles E. Beveridge. Baltimore: Johns Hopkins University Press, 1977.   M. H. S.

———. *Papers*. Vol. 3: *Creating Central Park*. Edited by Charles Capen McLaughlin, Charles E. Beveridge, and David Schuyler. Baltimore: Johns Hopkins University Press, 1983.   M. H. S.

Palaiseul, Jean. *Grandmother's Secrets: Green Guide to Health from Plants*. New York: Putnam, 1974.   James M. Smith

Parkinson, John. *Paradisi in Sole, Paradisus Terrestris* 1629; London: Methuen, 1904.   A.

Parkman, Francis. *The Book of Roses*. Boston: Tilton, 1866.   A.

Parsons, Samuel B. *Parsons on the Rose*. 1883, rpt. New York: Stanfordville, 1979.   A.

Peattie, Donald Culross. *Green Laurels: Lives of the Great Naturalists.* New York: Simon and Schuster, 1936.   M. H. S.

Pennell, Francis W. "Botanical Collectors of the Philadelphia Local Area." *Bartonia,* May 1942.   H. B. L.

Pickston, Margaret. *The Language of Flowers.* Facsimile, handwritten. London: Michael Joseph, 1968.   A.

Pinchot, Gifford, Consulting Forester. *Biltmore Forest* (Property of George W. Vanderbilt). Chicago: Riverside Press, 1893, for the World's Fair. M. H. S.

Platt, Charles A. *Italian Gardens.* New York: Harper, 1894.   M. H. S.

Porcher, Francis Peyre. *Resources of the Southern Fields and Forests: Medical Botany of the Southern States.* Charleston: Walker, Evans and Cogswell, 1869.   A.

Pursh, Frederick. *Flora Americae Septentrionalis.* Vol. 1. 2nd ed. London, 1816.   A.

Rehder, Alfred. *A Manual of Cultivated Trees and Shrubs.* 2nd ed. New York: Macmillan, 1974.   A.

Remini, Robert B. *Andrew Jackson and the Course of Empire.* Vol. 1. New York: Harper and Row, 1977.   I. P. L.

Repton, Humphry. *The Art of Landscape Gardening.* Edited by John Nolen. Boston: Houghton Mifflin, 1907.   A.

————. *Fragments of the Theory and Practice of Landscape Gardening.* London, 1816.   H. B. L.

Robbins, Christine Chapman. *David Hosack, Citizen of New York.* Philadelphia: American Philosophical Society, 1964.   M. H. S.

Robinson, W. *Garden Design.* London: John Murray, 1892.   T. Kane

Rogers, Andrew Denore, III. *American Botany, 1873–1892: Decades of Transition.* Princeton University Press, 1944.   M. H. S.

Rousseau, J. J. *Letters on the Elements of Botany.* Translated by Thomas Martyn. Addressed to a Lady. 6th ed. (Linnaean System). London, 1802.   A.

Sachs, Julius von. *History of Botany (1530–1850).* 1890; rpt. New York: Russell & Russell, 1967.   A.

Sargent, Charles Sprague. *Manual of North American Trees.* Boston: Houghton Mifflin, 1905; Dover Reprint, 1965.   A.

Sargent, Henry Winthrop. *Supplement to the Sixth Edition of "Landscape Gardening" by A. J. Downing.* Facsimile. New York: Funk and Wagnalls, 1859.   P. H. S.

Sayers, Edward. *Landscape and Ornamental Gardener: The American Flower Garden Companion.* Boston: Joseph Breck, 1838.   A.

Schlesinger, Arthur M., Jr. *The Age of Jackson.* Boston: Little, Brown, 1945.   I. P. L.

Schmidt, Carl F. *Fences, Gates, and Garden Houses.* Rochester, 1963.

Scott, Frank J. *The Art of Beautifying Suburban Home Grounds.* New York: Appleton, 1870.   A.

Sedgwick, Mabel Cabot. *The Garden Month by Month.* New York: Stokes, 1907.   A.

Simcoe, Mrs. J. G. *Diary.* Edited by Mary Quale Innes. Toronto: Macmillan, 1965.   A.

———. *The Diary of Mrs. John Graves Simcoe* (wife of the first lieutenant governor of the Province of Upper Canada, 1792–96). Edited by J. Ross Robertson. Toronto: William Briggs, 1911.

Sloan, Samuel. *Victorian Buildings.* 2 vols. Originally entitled *The Model Architect.* 1851. Dover Reprint, 1980.   A.

Smith, A. W. *A Gardener's Book of Plant Names.* New York: Harper and Row, 1963.   A.

Smith, Charles H. J. *Landscape Gardening; or, Parks and Pleasure Grounds.* Notes and additions by Lewis F. Allen. New York: C. M. Saxton, 1856.   A.

Stearn, W. T. "From Medieval Park to Modern Arboretum: The Arnold Arboretum and Its Historical Background," *Arnoldia* 32, no. 5 (1972): 173–97.   A.

Stephenson, John, and Churchill, James Morse. *Medical Botany: Medicinal Plants of the London, Edinburgh, and Dublin Pharmocopoeias.* 4 vols. London: John Churchill, 1831.   Sedgwick Library

Stevenson, Elizabeth. *Parkmaker: A Life of Frederick Law Olmsted.* New York: Macmillan, 1977.   H. H.

Stroud, Dorothy. *Humphry Repton.* London: Country Life, 1962.   A.

Taylor, Geoffrey. *Some Nineteenth Century Gardeners.* Stellington: Anchor Press, 1957. Tip Tree.   M. H. S.

Thomson, John. *Materia Medica,* pp. 605–831. 1832.   A.

Torrey, John. *Plantae Fremontiae: Description of Plants Collected by Col. J. C. Fremont in California.* Vol. 6. Smithsonian Contributions to Knowledge, 1854.   A.

Trollope, Frances. *Domestic Manners of the Americans.* 1832; Rpt. New York: Dodd, Mead, 1927.   A.

Upham, Charles Wentworth. *Life, Explorations, and Public Services of John Charles Fremont.* Boston: Ticknor and Fields, 1856.   A.

Van Ravenswaay, Charles. *A Nineteenth Century Garden.* New York: Main Street Press, 1977.   A.

Vaux, Calvert. *Villas and Cottages: A Series of Designs Prepared for Execution in the United States.* Dover Reprint, 1970.   A.

Washburn and Co. *Amateur Cultivator's Guide to the Flower and Kitchen Garden.* Boston: Washburn and Co., Seed Merchants, 1869.   A.

Waterman, Catherine H. *Flora's Lexicon: an Interpretation of the Language and Sentiment of Flowers, with an Outline of Botany and a Poetical Introduction*. Philadelphia: Hooker and Claxton, 1839; Herman Hooker, 1840.   A.

Williams, Harry T. *Window Gardening: The Culture of Flowers and Ornamental Plants for Indoor Use and Parlor Arrangement*. New York: Harry T. Williams, 1872.   A.

Wood, Alphonso. *Class-Book of Botany for Colleges, Academies, and other Seminaries*. 2nd ed. Claremont, N.H., 1848.   A.

Wood, George B., and Bache, Franklin. *The Dispensatory of the United States of America*. 16th ed. Philadelphia: Lippincott, 1891.   A.

Wright, C. H., and Demar, D. *Johnson's Gardener's Dictionary*. New ed., revised and enlarged. London: George Bell, 1894.   A.

Youmens, Eliza A. *First Book of Botany: Designed to Cultivate the Observing Powers of Children*. New York: Appleton, 1870.   A.

Zaitzevsky, Cynthia. *Frederick Law Olmsted and the Boston Park System*. Cambridge: Harvard University Press, 1982.   A.

Index

This index includes chiefly the people mentioned and the places important to them. The Appendix stands as its own index.